CLIMATE IMPACT A
ADAPTATION ASSESSMENT

A Guide to the IPCC Approach

Martin Parry

and

Timothy Carter

earthscan
from Routledge

First published by Earthscan in the UK and USA in 1998

For a full list of publications please contact:

Earthscan
2 Park Square, Milton Park, Abingdon, Oxon 0X14 4RN
Simultaneously published in the USA and Canada by Earthscan
711 Third Avenue, New York, NY 10017

Earthscan is an imprint of the Taylor & Francis Group, an informa business

A catalogue record for this book is available from the British Library

ISBN: 978-1-85383-266-6 (pbk)

Typesetting and page design by PCS Mapping & DTP, Newcastle upon Tyne

Cover design by Andrew Corbett

CONTENTS

LIST OF FIGURES, TABLES AND BOXES

FIGURES

TABLES

BOXES

ACRONYMS, ABBREVIATIONS AND CHEMICAL FORMULAE

ASLR	accelerated sea-level rise
BLS	Basic Linked System
BP	before present
CCIRG	United Kingdom Climate Change Impacts Review Group
CFC	chlorofluorocarbon
CGE	computable general equilibrium (models)
CH_4	methane
CO_2	carbon dioxide
CRU	Climate Research Unit, Norwich, UK
EASM	Egyptian Agricultural Sector Model
ENSO	El Niño/Southern Oscillation
EPA	Environmental Protection Agency (US)
ET	evapotranspiration
FAO	Food and Agriculture Organization
GCM	general circulation model
GDP	gross domestic product
GE	general equilibrium
GFDL	Geophysical Fluid Dynamics Laboratory
GHG	greenhouse gas
GIS	geographical information systems
GISS	Goddard Institute for Space Studies
GNP	gross national product
GRID	Global Resource Information Database (UNEP)
HEM	Harmonization of Environmental Monitoring
IFIAS	International Federation of Institutes for Advanced Study
IBSNAT/	International Benchmark Sites Network for Agrotechnology
ICASA	Transfer – International Consortium for Application of Systems Approaches to Agriculture
IIASA	International Institute for Applied Systems Analysis
IMF	International Monetary Fund
IO	input–output
IPCC	Intergovernmental Panel on Climate Change
ISRIC	International Soil Reference and Information Center
LDC	Less developed country
LP	linear programming
MAGICC	Model for the Assessment of Greenhouse Gas-Induced Climate Change

MINK	Missouri, Iowa, Nebraska and Kansas study on the US corn belt
MIT	Massachusetts Institute of Technology
mmt	million metric tons
NASA	National Aeronautics and Space Administration, US
N_2O	nitrous oxide
NCAR	National Center of Atmospheric Research, Boulder, Co, US
NDU	National Defense University
NOAA	National Oceanographic and Atmospheric Association
OECD	Organization for Economic Cooperation and Development
PAR	photosynthetically active radiation
PDSI	Palmer Drought Severity Index
ppb	parts per billion
ppbv	parts per billion by volume
ppm	parts per million
ppmv	parts per million by volume
SO_2	sulphur dioxide
SO_x	sulphur oxides
UKMO	United Kingdom Meteorological Office
UNEP	United Nations Environment Programme
UNESCO	United Nations Educational, Scientific and Cultural Organization
UNFCCC	United Nations Framework Convention on Climate Change
WCIRP	World Climate Impact Assessment and Response Strategies Studies Programme
WEC	World Energy Council
WHO	World Health Organization
WMO	World Meteorological Organization
WRI	World Resources Institute

PREFACE

Our aim in writing this book has been to provide a readable guide to the approach toward climate impact and adaptation assessment which has been adopted by the Intergovernmental Panel on Climate Change (IPCC) and reported in the IPCC *Technical Guidelines for Assessing Climate Change Impacts and Adaptations* (Carter et al, 1994). As lead authors of the Guidelines we were aware of the many omissions that were necessary in our requirement to condense the information into less than 60 pages. This is an expansion of that information; it contains more background on how the approach has been developed over the years, and more examples of its use. We wish to emphasise that it is not, nor was ever intended to be, an IPCC document. We have, however, kept closely to the IPCC approach and readers can be sure that our aim has been solely one of illustration. With this aim we inevitably have drawn upon the work of many scientists – particularly the 48 experts and 26 reviewers who worked with us in drafting the IPCC *Guidelines*. Some, however, deserve special mention: Hideo Harasawa and Shuzo Nishioka, who were our fellow lead authors; Renate Christ, Paul Epstein, Eugene Stakhiv and Joel Scheraga, who were co-authors; Sue Lane and Ruth Zivanovic, who keyboarded some of the text; and Cynthia Parry, who edited, proofread and indexed the text.

Martin Parry
Timothy Carter
November 1997

Chapter 1

INTRODUCTION

THE SCOPE OF THIS BOOK

The purpose of this book is to provide a guide to current methods of assessing the impacts from, and potential adaptations to, changes of climate. The nature of such changes may be various: for example, they may be future and long term, due to present and future emissions of greenhouse gases, or they may be shorter term, historical and unrelated to anthropogenic activity (such as the Little Ice Age in Europe in 1550–1750 or the Little Climatic Optimum which preceded it in about 1050–1250). While different methods of assessment may suit different types of climate change, the broad approach is the same.

This approach has been developed through the work of many scientists over the past 20 years and has recently been thoroughly examined and rewritten for the Intergovernmental Panel on Climate Change (IPCC).[1] This book provides a more complete explanation of the IPCC approach, includes a number of worked exam-

ples, and is written in a form suitable for the non-expert scientist. The structure of the book follows a seven step strategy approved by the IPCC for climate impact and adaptation assessment:

1 definition of the problem;
2 selection of the method;
3 test of the method;
4 selection of scenarios;
5 assessment of impacts;
6 assessment of autonomous adjustments; and
7 evaluation of appropriate adaptation strategies (see Figure 1.1).

A description of each of these seven steps is given in each chapter of this book together with illustrations and reference to published examples elsewhere.

It should be emphasized that no single method is prescribed here, but the reader is introduced to an array of methods, some of which will be more suitable than others to the task in hand – be it a national level assessment of

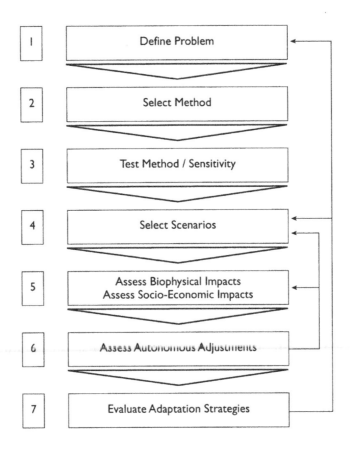

Note: Thin return arrows indicate steps that may be repeated.

Source: Adapted from Carter et al, 1994.[2]

Figure 1.1 *Seven steps of climate impact assessment, adopted by the IPCC*

adaptations to greenhouse gas-induced climate change or a small-scale study of the effects of drought over a relatively short time period. What is important is that the different methods are broadly comparable and thus offer a means by which compatible assessments may be made of impacts and adaptations in different regions or geographical areas, in different economic sectors and in countries at different levels of development. These methods will be particularly useful in enabling countries to meet, in part, their commitments under the UN Framework Convention on Climate Change (UNFCCC) which requires signatories to take steps to reduce impacts from greenhouse gas-induced climate change. The IPCC approach to climate impact and adaptation assessment is currently being developed into simplified workbooks which can be completed, sector by sector, to provide national assessments. The workbooks have recently been tested in a number of countries. The ultimate aim is to enable nation states to estimate, firstly, their contribution to greenhouse gas emissions (through use of the OECD/UNEP greenhouse gas emissions inventories) and, secondly, possible impacts due to greenhouse gas-induced climate change (through internationally agreed methods of impact assessment).

The purpose of climate impact assessment

As the next chapter shows, climate can change in several different ways. The physical factors causing change may be natural or anthropogenic, large scale (eg global) or small scale, long term or short term. The impacts implied by such changes and the adaptations that may be required are also likely to vary greatly, and over the past 20 to 30 years scientists have developed a variety of methods to assess them. Although these methods vary quite widely they have sufficient features in common to enable us to address them as a single broad approach – an approach we term *climate impact assessment*.

The goal of climate impact assessment is to ensure optimal use of available climatic resources through, firstly, the measurement of impacts (both positive and negative) that are likely to stem from either short-term variations or long-term changes of climate and, secondly, the evaluation of the array of adaptations likely to represent the optimal response to such impacts (by taking advantage of the positive impacts and minimizing the negative ones). Within this overall goal there may be a number of more specific objectives, such as the identification of areas, populations or regions most sensitive to climate change, the reduction of vulnerabilities in these regions through improved resource management, and enhanced long-range planning by, for instance, forecasting future climate impacts due to drought. To achieve these objectives climate impact assessment frequently follows the sequence of tasks shown in Figure 1.1: firstly, to establish the scientific means of effective assessment and then to communicate results of the assessments to policy-makers.

Policy-makers require climate impact assessments to provide them with the necessary scientific information for policy decisions. These decisions include considering the options for mitigating climate change or those for adapting to it, either by coping with, ameliorating or exploiting its projected impacts. Assessments are required for different time and space scales, reflecting the time horizons and areas to which planning and decision-making apply.

Climate impact assessment must address an inherently global phenomenon affecting all nations, so it is desirable that assessments are conducted in a transparent manner, with comparable assumptions and internally consistent procedures. Comparability among assessments is of great importance in appraising

3

the range of appropriate response actions at the international, national and regional levels. Decision-makers must have confidence that, at a minimum, the basic assumptions are uniform (such as use of a common set of scenarios), that the various models and analytical tools are used correctly, and that the evaluation of impacts properly takes into account future impacts due to socio-economic and technological changes that would occur even in the absence of climate change. These are persuasive arguments for adopting a standardized approach to climate impact assessment. In the chapters that follow, each step in the assessment sequence is explained and illustrated, providing an expanded guide to the standard approach developed and adopted by the IPCC.

Chapter 2

UNDERSTANDING AND PREDICTING CLIMATE CHANGE

DEFINITIONS

The earth's atmosphere, a thin layer of gases that surrounds the globe, both protects and sustains life on earth. The atmosphere is continually in motion, driven by radiation energy from the sun, differences in the heating of the earth's surface between the poles and equator, and the earth's rotation itself. The state of the atmosphere – its temperature, moisture content, pressure and airflow – is usually described in terms of weather and climate.

Weather refers to the local state of the atmosphere during time periods ranging from minutes to days. Weather events, such as rainfall, heatwaves, gales, frost and snowfall, are familiar to most of us because they affect our everyday lives. They arise from weather systems: instabilities in the atmosphere which move, develop, mature and decay.

Climate, on the other hand, can be thought of as an average of the weather over a period of years or decades. It describes the characteris-

tic weather conditions to be expected in a region at a given time of year, based on long-term experience. By international convention, weather observations are commonly averaged over a period of 30 years to produce the statistics that describe the climate, such as the mean, variability and extremes.

Climate, like the weather, is also variable. The climate of a region is never identical from one year to the next: the frequency of certain types of weather events varies from year to year and decade to decade. For example, one year may bring heavy rainfall events and flooding, the next may offer prolonged drought. Such short-term variations are usually referred to as *climatic variability*.

Climate also changes over longer time spans, ranging from decades to hundreds of thousands of years. A *climate change* can be defined as a change in the average climate (or its variability) from one averaging period to the next. Some changes in climate can be large and dramatic, as is clearly shown

Box 2.1
THE EL NIÑO/SOUTHERN OSCILLATION PHENOMENON

Some of the most productive fishing grounds in the world are to be found in the Pacific Ocean off the west coast of Peru and Ecuador in South America. Here the ocean currents bring cool nutrient-rich waters upwelled from lower levels. In the first few months of each year, a warm southward current usually modifies the cool waters. However, every two to seven years this warming starts earlier, in December, and is far stronger, sometimes lasting as long as a year or two. This event, labelled El Niño ('the Christ child') because of its frequent appearance around Christmas time, can devastate local fisheries and often leads to torrential rains over the usually arid coastal strip.

Although once thought to be a localized event, the El Niño phenomenon and its counterpart, La Niña (a cooling of the eastern tropical Pacific), have been linked

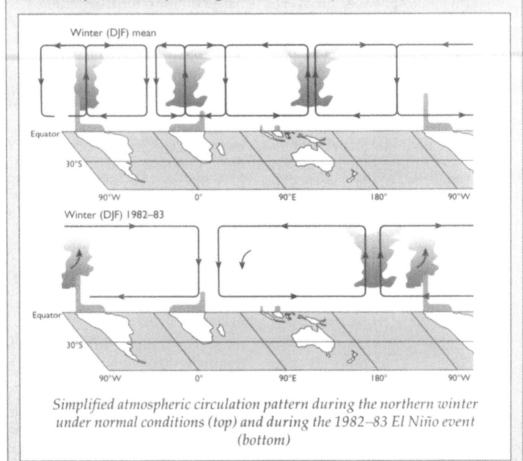

Simplified atmospheric circulation pattern during the northern winter under normal conditions (top) and during the 1982–83 El Niño event (bottom)

to large-scale variations in oceanic temperature that affect most of the tropical Pacific. Moreover, El Niño/La Niña are usually accompanied by a change in the atmospheric circulation known as the Southern Oscillation. This refers to a periodic reversal of the gradient of atmospheric pressure between the eastern and western Pacific, conventionally measured between Darwin in Australia and the South Pacific island of Tahiti. Together, the El Niño/Southern Oscillation (ENSO) phenomenon is now recognized as one of the main sources of interannual variability in climate around the world.

The effects of an ENSO event are sometimes felt far away from the tropical Pacific. This is because of 'teleconnections', physical relationships occurring over long distances due to the dynamics of atmospheric and oceanic waves. Characteristic effects include enhanced rainfall over the central Pacific, Peru, Ecuador and the southern United States and drought in Australia, north-eastern Brazil and southern Africa (see figure). The economic cost of the severe 1982–83 El Niño event is thought to have totalled more than US$8 billion worldwide. It is also estimated that global mean temperatures increase by as much as 0.3°C in the months after a strong ENSO event.[1] This is thought to be due to increased heat transport from the tropics to higher latitudes and latent heat release into the atmosphere related to evaporation processes.

from reconstructions of temperature changes between the ice ages and warmer interglacial periods (see below).

Some of these fluctuations in climate are due to processes that are internal to the climate system, reflecting interactions between the atmosphere, oceans, sea ice, and features of the land surface such as snow and ice cover, vegetation and surface water. Other fluctuations are due to external forcing, for example from changes in the radiation emitted by the sun, slow changes in the earth's rotation and orbit, movements of the land relative to the oceans, and changes in the composition of the atmosphere and ocean.

Most of these forcing factors are regarded as natural in origin. For example, one of the strongest natural, internal climatic fluctuations, with an irregular time interval of several years, is the El Niño/Southern Oscillation (ENSO) phenomenon (see Box 2.1). This influences weather patterns over large areas of the tropics and beyond. Examples of natural external forcings include the earth's orbital variations, which are thought to be responsible for the alternating glacial and interglacial cycles over hundreds of thousands of years, and major volcanic eruptions, which can inject large quantities of material into the atmosphere and may affect global temperatures over periods of years.

However, some external forcings may occur as a result of human activities, such as increases in trace gases and aerosols which are associated with the enhanced greenhouse effect. Human-induced changes in climate and their possible future impacts form a major focus of this book and are considered in more detail below. First, however, it is instructive to examine past changes in climate. These changes can sometimes be related to recorded impacts on human societies or ecosystems. They also provide evidence for recent and ongoing climatic trends.

CLIMATE CHANGES IN THE PAST

Past climate changes can be separated into three categories, distinguished by the time periods they represent and by the sources of information used for climatic reconstruction: palaeoclimatic changes, historical changes and changes in the instrumental period.

Palaeoclimatic changes

Palaeoclimatic changes refer to all climate changes occurring before the historical period. For convenience, this can be regarded as the period before about 1000 years ago (before present, or BP). Since there were no direct instrumental measurements of the climate in the distant past, and without surviving eye witness accounts, climatologists must make use of indirect 'proxy' evidence to establish a chronology of climate changes. Some of the main sources of

proxy information, along with the timescales they can represent, are given in Table 2.1. Using these and some other sources of evidence, it is known that a period of glacial–interglacial cycles (the Pleistocene), when climate was mainly cooler than at present, began about two million BP. Six or seven ice ages can be identified at roughly 100,000-year intervals, with warmer interglacials in between. This regular cycle of glaciations can be related directly to predictable orbital variations of the earth.

Orbital variations can also be calculated into the future and indicate that a change will occur over the next 50,000 years towards conditions previously associated with glacial periods.[2] However, the time frame of these changes is very different from the decadal or century-scale warming anticipated under the enhanced greenhouse effect (see below).

Ice cores from Greenland reveal that temperature changes at the end of the last glaciation may have been extremely rapid. Over Greenland, a warming of some 7°C may have occurred in a matter of a few decades around 11,500 BP.[3] Thus, the rate of global climate change anticipated during future decades due to the contemporary build-up of greenhouse gases in the atmosphere (see below) is not unprecedented, at least within the time frame of the Pleistocene ice ages. On the other hand, the same Greenland ice cores also indicate that mean annual temperatures during the last 10,000 years (the Holocene period) have been

Table 2.1 *Sources of evidence for reconstructing the past climate*

Source	Time resolution	Start of record (BP)	Region covered
Deep sea cores/ fossil foraminifera	100–2500 years	2 million years+	Samples from all oceans
Fossil flora and fauna	100 years	125,000 years (continuous) 2 million years+ (fragmented)	Land (especially Eurasia, North America)
Ice cores	1–10 years	200,000 years	Antarctica, Greenland
Glaciers (margins)	Depends on dating method	17,000–22,000 years	Most mountain regions and high-latitude lowlands
Lake sediments	1 year	10,000 years	Europe, North America, Japan
Tree rings	1 year and less	6000–8000 years	Eurasia, North America, temperate southern hemisphere
River and lake levels	1 day – 1 year	2000 years/fragments up to 5000 years	Eurasia, Africa, Americas
Annals, reports and chronicles	1 day – 1 season	1000 years/scattered samples from 2000+ years	Eurasia
Weather registers and diaries	1 day	700 years	Europe
Standard meteorological instruments	Almost instantaneous	350 years (surface)/ 70 years (upper air)/ 150 years (ship-borne)	Longest in Europe; most recent in Antarctica and over the southern ocean

Source: Lamb, 1982.[4]

remarkably stable. It is thought unlikely that global mean temperatures varied by more than 1°C per century during this period.[5] Nevertheless, regional changes have occurred, and pollen reconstructions indicate that summer temperatures in the period between about 5000 and 6000 BP were higher than the present-day in many middle and high-latitude regions of the northern hemisphere.[6]

Annual precipitation was also higher than present over high latitudes and at lower latitudes over central Asia, the Sahara and the Middle East, where summer temperatures may have been cooler than present. Indeed, there is abundant evidence of human activities such as hunting and fishing in the present Sahara, which is today one of the most arid regions on earth. The altered patterns of precipitation over Europe and north Africa are consistent with a northward shift of the circumpolar westerly airstreams to northern Europe and a corresponding northward shift of the high-pressure systems that currently maintain stable, dry conditions over north Africa, allowing the northward penetration of summer monsoon rains into the Sahelian zone and the southern Sahara.[7]

Historical changes (the last 1000 years)

The historical period, defined here as the last 1000 years, provides not only a more detailed chronology of regional climatic variations, but also documentary evidence for the impacts these variations may have had. Two periods during this millennium have received special attention as presumed warm and cool periods globally: the Medieval Warm Period and the Little Ice Age. However, much of the evidence for these has been obtained from data-rich regions of the northern hemisphere, and recent evidence from other regions has cast some doubt on their global applicability. Lack of data precludes the large-scale reconstruction of climate during the medieval period (from the 9th to 14th centuries AD), but while there is clear evidence for a period of warmth in parts of Europe in the 11th and 12th centuries, other records either show no evidence, or the timing of the warm period was different (Figure 2.1).

Similarly, the Little Ice Age, for long thought to be a worldwide cool period of some 400 to 500 years, is now being reinterpreted as a period of both cool and warm anomalies that varied in importance geographically. In the northern hemisphere, a composite reconstruction from various sources indicates that the coldest summers were experienced from 1570 to 1730 and during most of the 19th century. However, warmer conditions were common in the early 16th century and during much of the 18th century.[8]

Overall, the evidence suggests that the 20th century has been at least as warm as any century since 1400, and the warming since the late 19th century appears unprecedented, at least over much of the northern hemisphere.[9] A 20th-century warming is also evident in records from Australasia and parts of South America, but there are also exceptions in both hemispheres to these generalized trends.

Changes during the instrumental period

The most accurate determinations of variations in global climate are based on instrumental measurements with standard equipment at sites around the world, both on land and at sea, supplemented in recent years with monitoring from satellites outside the earth's atmosphere. There are sufficient observations of surface temperature to allow reconstructions of global mean annual temperature changes over the past 130 years (see

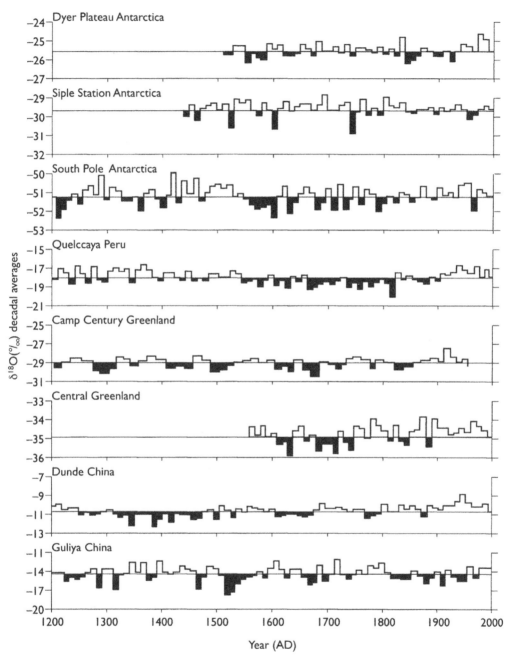

Note: Shaded areas are cooler than the mean for each record.

Source: Adapted from Thompson et al, 1993.[10]

Figure 2.1 *Decadal averages of oxygen isotope records from ice cores at different locations*

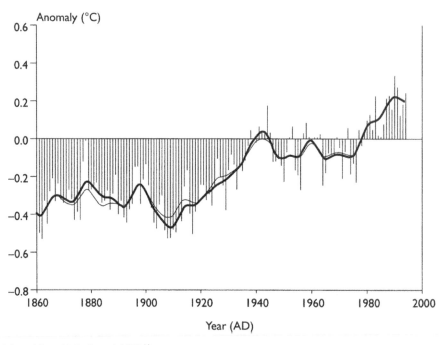

Source: Adapted from Nicholls et al, 1996.[11]

Figure 2.2 *Combined global mean annual land-surface and sea-surface temperature anomalies (°C) 1861–1994 relative to the 1961–90 mean*

Figure 2.2). Confirming the warming trends found in the proxy data described above, the instrumental record shows a warming of between 0.3 and 0.6°C since the late 19th century, with a warming of 0.2 to 0.3°C over the 40 years up to 1994, the period with the most reliable data.[12] Years during the 1980s and 1990s have been amongst the warmest since 1860. The warming has not been uniform, however, but is strongest over the continents between 40°N and 70°N. A few areas, such as the North Atlantic Ocean north of 30°N and nearby land areas, have cooled in recent decades. Over many land areas, there has also been a trend since the 1950s towards a reduced daily temperature range, meaning that where average temperatures have increased, nights have warmed by more than days.[13] This is thought to be due to increased cloudiness and possibly also to increases in atmospheric aerosols due to fossil fuel burning.

Precipitation averaged over land areas increased from the start of the century up to about the 1960s, but has decreased since about 1980.[14] There are large regional variations in these changes and these are further complicated by ENSO events. Prominent among the regional changes are a long-term increase in winter precipitation at high latitudes in the northern hemisphere and a sharp decrease in precipitation from the 1960s over the subtropics and tropics, from Africa to Indonesia. This includes the Sahelian zone of west Africa, which has recorded only about half the precipitation in recent decades compared to the wet 1950s. However, similar dry periods in this region have also occurred in the historical and recent geological past.[15]

There is increasing confidence among climatologists that the main features of the observed record of global climate temperature change during the past century can be explained, although details remain that are not fully understood. Some short-term variations in the order of several tenths of a degree Celsius can been attributed to natural causes, such as the ENSO phenomenon (producing surface warming), volcanic eruptions (surface cooling), changes in atmospheric circulation patterns or minor variations in radiation output from the sun. However, in an authoritative statement agreed by scientists in 1995, it is now accepted that natural causes alone are not sufficient to explain the rate and pattern of long-term change during the past century, but that the evidence points towards 'a discernible human influence on global climate'.[16] This refers to the observed build-up of greenhouse gases in the atmosphere since preindustrial times due to human activities, especially fossil fuel combustion and land use changes.

GREENHOUSE GAS-INDUCED CLIMATE CHANGE

The main energy source warming the earth is radiation from the sun. This is shortwave radiation, due to the high temperature of the sun (6000°C), and includes visible light and ultra-violet radiation. The earth intercepts some of this radiation energy, which warms the surface and is re-emitted to balance the incoming energy. The earth is much cooler than the sun, and it emits terrestrial radiation at longer wavelengths in the infra-red part of the spectrum, invisible to the human eye.

The earth's atmosphere is made up of gases, some of which are able to absorb radiation at particular wavelengths. The gases that make up the bulk of the atmosphere, nitrogen (78 per cent) and oxygen (21 per cent), neither emit nor absorb terrestrial radiation. However, water vapour, carbon dioxide and a number of other trace gases absorb some of the terrestrial radiation leaving the earth, while being transparent to the incoming solar radiation. The effect of this is to warm the earth's surface to an average temperature of 15°C, which is highly amenable for life on earth and is some 34°C higher than the frigid temperature (−19°C) to be expected in the absence of an atmosphere.[17] This is the *natural greenhouse effect*, so called because the absorbing gases act like the glass in a greenhouse, allowing the sun's rays to penetrate inside, but re-emitting some of the outgoing terrestrial radiation back into the greenhouse, thus helping to keep it warm.

Over the last two centuries, continuous increases in atmospheric concentrations of many of these 'greenhouse gases' have been observed. This period coincides with the industrialization of human societies, beginning in Europe but spreading to become the global phenomenon it is today. Industrialization has been associated with massive land clearance for agriculture and the combustion of fossil energy sources such as coal, oil and gas. Both of these activities have resulted in increasing emissions of natural greenhouse gases such as carbon dioxide, methane and nitrous oxide into the atmosphere.

Growing emissions of greenhouse gases have resulted in increasing atmospheric concentrations since pre-industrial times (see Table 2.2). The gases are well mixed in the atmosphere and their global average concentrations can be estimated using ice core evidence and more recently by direct measurements. Rising greenhouse gas concentrations mean that more terrestrial radiation is absorbed and re-emitted back to earth, resulting in a warming of the atmosphere near the surface. This is known as the *enhanced greenhouse effect*. The effectiveness of a greenhouse gas in warming the atmosphere (its 'radiative forcing') depends both on its concentration and on the amount of time it remains in the atmosphere. These are both shown in Table 2.2.

In addition to the natural greenhouse gases, new chemical compounds, the halocarbons, have been developed for use in foam packaging and as aerosol propellants. When released into the atmosphere, halocarbons can reside there for decades or even centuries. As well as being effective greenhouse gases, some halocarbons are also known to be responsible for depleting ozone in the upper atmosphere (stratosphere) over polar regions: forming the so-called 'ozone hole'. Ozone protects life on the surface from dangerous levels of ultraviolet radiation and concern about ozone depletion led to the signing in 1989 of an international agreement, the Montreal Protocol, to eradicate the use of the ozone-destroying chlorofluorocarbons (CFCs). This protocol, and subsequent amendments, has already seen a rapid decline in CFC emissions and their replacement with other compounds that are benign for the ozone layer. However, these replacements, like CFCs, are also greenhouse gases (see Table 2.2).

As well as the gases shown in Table 2.2, other constituents in the atmosphere may also contribute to radiative forcing but are more localized, variable and thus difficult to measure. One of these is ozone in the lower atmosphere (troposphere), which is a byproduct of fossil fuel combustion and can often be found in high concentrations over urban areas during stable weather conditions. Unlike its protective role in the stratosphere, ozone near the surface is both a green-

Table 2.2 *Greenhouse gas concentrations, increases and lifetimes*

	CO_2	CH_4	N_2O	CFC-11	HCFC-22 (a CFC substitute)
Pre-industrial concentration	~280 ppmv	~700 ppbv	~275 ppbv	zero	zero
Concentration in 1994	358 ppmv	1720 ppbv	312 ppbv	268 pptv	110 pptv
Rate of concentration change	1.5 ppmv/yr 0.4%/yr	10 ppbv/yr 0.6%/yr	0.8 ppbv/yr 0.25%/yr	0 pptv/yr 0%/yr	5 pptv/yr 5%/yr
Atmospheric lifetime (years)	50–200	12	120	50	12

Source: IPCC, 1996.[18]

house gas and a toxic pollutant for living organisms. Where they have been measured, ozone concentrations also appear to be increasing in many regions of the world. Another important factor affecting radiative forcing is atmospheric aerosols. These are small particles or droplets that can be produced both naturally (for example, by dust storms and volcanic activity) and anthropogenically (through fossil fuel and biomass burning). Their concentrations vary regionally, with the largest concentrations around industrial regions, especially in the developed world. They only remain in the lower atmosphere for a few days before they are washed out, but can reside for several years in the stratosphere. Aerosols influence the earth's radiation balance both directly, by scattering and absorbing solar radiation, and indirectly, by altering the properties and lifetime of clouds. Apart from soot particles, which slightly warm the surface, the net effect of aerosols is to cool the surface

– a negative radiative forcing.

The relative contributions of greenhouse gases and aerosols to radiative forcing since preindustrial times, and the confidence attached to the estimates, are summarized in Figure 2.3. Of the well-mixed greenhouse gases, carbon dioxide has contributed about 64 per cent of the radiative forcing, methane about 19 per cent, halocarbons 11 per cent and nitrous oxide 6 per cent.[19] Also shown are estimates of the radiative forcing due to reductions in stratospheric ozone (slightly negative) and the possible effects of natural variations in solar output (thought to be positive). In spite of the uncertainties surrounding estimates of regional aerosol effects, the net forcing over most regions since preindustrial times is clearly positive, and according to the theory of the enhanced greenhouse effect, it should have caused a warming of the earth's surface. However, the amount and geographical pattern of this warming depends

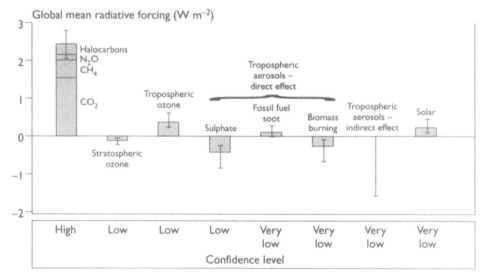

Note: Estimates of the uncertainty range are shown by error bars.

Source: Adapted from Schimel et al, 1996.[20]

Figure 2.3 *Estimates of globally and annually averaged anthropogenic radiative forcing (Watts per square metre) due to changes in greenhouse gases and aerosols from pre-industrial times to the present day and due to natural changes in solar output from 1850 to the present day*

on how the climate system responds to radiative forcing. This is considered in the following section.

MODELLING THE CLIMATE SYSTEM

The response of the climate system to radiative forcing is a complex problem involving interactions and feedbacks between the atmosphere, the oceans and features on the land surface. In order to obtain a better understanding of this problem it has become customary to use climate models, which are mathematical descriptions of the large-scale physical processes governing the climate system. The most detailed of these models are coupled ocean–atmosphere general circula-tion models (GCMs). These depict the climate using a three-dimensional grid over the globe, typically having a horizontal resolution of about 250 kilometres and 10 to 20 layers in the vertical. Many physical processes, such as those related to clouds, occur at smaller scales and cannot be prop-erly modelled. Instead, their known properties must be averaged over the larger scale in a technique known as parameterization.

An important test of GCMs is their ability to reproduce the present-day climate. The most recent models are fairly successful at simulating the important large-scale features of the climate system, including seasonal, geographical and vertical variations, but there are still discrepancies at the regional scale. GCMs have also been able to reproduce the global cooling following the eruption of Mt. Pinatubo in June 1991[21] and several features of regional circulation related to the ENSO phenomenon.[22]

The warming effect of greenhouse gas-induced radiative forcing is modified by climate feedbacks which either amplify (positive feedback) or dampen (negative feedback) the original response. Some of the main uncertainties in GCM estimates can be traced back to the simulation of these feedback mechanisms. The more important of these include:[23]

- the water vapour feedback, whereby a warmer atmosphere can hold more water vapour which, being a greenhouse gas, leads to enhanced warming, producing a positive feedback;
- the cloud–radiative feedback, which can be either negative or strongly positive, depending on the cloud height, thickness and radiative properties;
- ocean circulation, which transports large amounts of heat from the tropics to high latitudes and which may slow down regionally due to radiative forcing, thus acting as a local negative temperature feedback; the large heat capacity of the oceans also introduces a delay in the climate response to radiative forcing which varies regionally;
- ice and snow albedo feedback,

which refers to the strong reflection of solar radiation from snow and ice surfaces; as ice melts at the surface, less solar radiation is reflected, leading to further warming (a positive feedback).

These uncertainties become important when attempting to predict the global climate response to future radiative forcing. Thus, different GCMs may simulate quite different responses to the same forcing, simply because of the way certain processes and feedbacks are modelled. A common measure of uncertainty is the *climate sensitivity*, which can be defined as the change in global mean temperature for a radiative forcing equivalent to that obtained with a doubling of atmospheric CO_2. Many GCM runs have been made for this 2 x CO_2 atmosphere, and from these a range of uncertainty of 1.5 to 4.5°C is commonly accepted for the climate sensitivity, with a 'best estimate' of 2.5°C.[24]

ESTIMATES OF FUTURE CLIMATE

There is little doubt that the increases in greenhouse gas concentrations observed over the past century will continue well into the 21st century (see Table 2.2). We can be confident of this for two reasons. Firstly, as human population grows and fossil fuel combustion continues to rise, the rate of emissions of these gases into the atmosphere is unlikely to be abated for many decades. Secondly, even if emissions were reduced immediately, concentrations would continue to rise for some time because of the long residence times of these gases in the atmosphere and the slow uptake and release of these gases in the oceans.

In 1992, the Intergovernmental Panel on Climate Change (IPCC) developed six projections of future global emissions to the year 2100 – the IS92 scenarios.[25] These scenarios attempt to represent six possible paths of future emissions, based on alternative assumptions about world population growth, economic activity and energy use. They include all of the greenhouse gases as well as sulphate aerosols. They range from a low scenario (IS92c), which assumes low population growth, low economic growth and severe constraints on fossil fuel supplies, to a high scenario (IS92e), which assumes moderate population growth, high economic growth, high fossil fuel availability and a phase-out of nuclear power.

In order to estimate the greenhouse gas concentrations in the atmosphere resulting from these emissions scenarios, simple models can be used to represent the processes that transform and remove the different gases from the atmosphere. Figure 2.4a shows the modelled concentrations of CO_2, the most important greenhouse gas, under the six IS92 emissions scenarios. Even under the lowest emission scenario, IS92c, concentrations in 2100 are 35 per cent above 1990 levels and still growing.

Under the IS92e scenario, the corresponding increase is 170 per cent. For methane and nitrous oxide the respective ranges are 22 per cent to 175 per cent and 26 per cent to 40 per cent.[26] The scenarios of future greenhouse gas concentrations, along with projections of aerosol concentrations based on the emissions scenarios, can be converted to radiative forcing scenarios using knowledge about the radiative properties of each constituent. These are shown in Figure 2.4b, which also includes the forcing estimated from 1765 to 1990.

Two alternative approaches are commonly adopted to estimate the response of the climate system to radiative forcing. The first, more detailed, approach involves the use of GCMs. However, because of the complexity of these models and the large resources required to conduct model runs, only a limited number of simulations can be conducted. A second approach, therefore, is to use simpler upwelling diffusion/energy balance models, which generalize the results of the GCMs but only produce estimates of global mean temperature change. These models allow scientists to explore the likely mean temperature response to alternative forcing scenarios, assuming different levels of climate sensitivity.

Figure 2.5 shows how different assumptions can be combined to produce a range of global temperature projections. Two sets of curves are shown. The dashed curves represent the modelled temperature response to

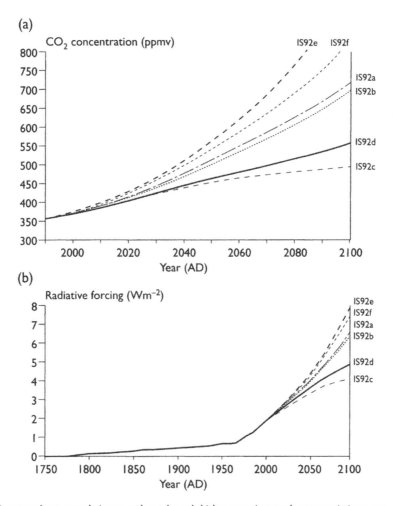

Note: IS92 scenarios: **a** moderate population growth, moderately high economic growth, some emissions controls; **b** as **a** but stricter emissions controls; **c** low population and economic growth, severe constraints on fossil fuel; **d** similar to **c** but higher economic growth; **e** similar to **a** but higher economic growth and phase-out of nuclear power; **f** similar to **a** but higher population growth.

Source: Adapted from IPCC, 1996.[27]

Figure 2.4 (a) *Estimated atmospheric CO_2 concentrations, 1990–2100, and (b) estimated global and annual average radiative forcing from 1765 to 1990, and projected to 2100 under the six IS92 emission scenarios*

forcing by greenhouse gases only. The solid curves are estimates of the response to greenhouse gases and aerosols combined. In each case, the top curve shows global mean annual temperatures up to the year 2100, assuming both the highest emission scenario (IS92e) and the highest climate sensitivity (4.5°C). The bottom curve depicts the temperature

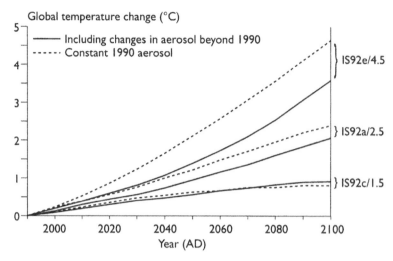

Note: See text for explanation.

Source: IPCC, 1996.[28]

Figure 2.5 *Projected global mean surface-temperature change extremes and central estimate from 1990 to 2100, based on different assumptions*

response for low emissions (IS92c) and low climate sensitivity (1.5°C). The middle curve is a central estimate, combining the IS92a emissions scenario and the best estimate climate sensitivity (2.5°C). Estimates for greenhouse gas forcing alone are shown because the projections of future aerosol concentrations and their effect on temperature are still highly uncertain. Together, these curves imply that global mean temperature by the year 2100 could range between about 0.8°C and 4.5°C above 1990 levels.

Projections of global mean annual temperature change can be quite useful for examining policy issues such as the effect of reducing greenhouse gas emissions on global climate. However, they tell little about the magnitude, rate and pattern of climate change in different regions. This information is of critical importance for assessing the impacts of climate change on ecosystems, water resources and economic activities such as agriculture and forestry, all of which are highly region-specific. The main source of information on possible regional changes in climate is GCMs. Typical results from a coupled ocean–atmosphere GCM are presented in Figure 2.6. They depict the mean annual temperature and precipitation change by the period 2040–2070 relative to 1960–1990 under a radiative forcing scenario for greenhouse gases only, based on the IS92a emissions scenario described above.

The main results from this and other similar simulations include:[29]

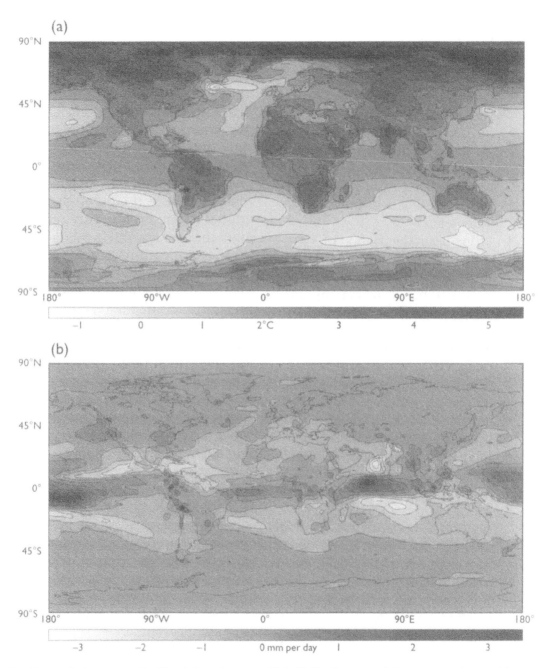

Note: Estimated using an ensemble of four independent runs with the Hadley Centre coupled ocean–atmosphere model HADCM2* for changes in greenhouse gas concentrations approximating the IS92a emissions scenario.**

Source: * Johns et al, 1997;[30] ** Hadley Centre, 1997.[31]

Figure 2.6 (a) *Mean annual temperature change (°C) and (b) Annual precipitation change (mm per day) between 1960–90 and 2040–70*

- greater surface warming of the land than the oceans in winter;
- a minimum warming around Antarctica and in the northern Atlantic associated with deep-water formation;
- maximum warming at high northern latitudes in late autumn and early winter associated with reduced sea ice and snow cover;
- little warming over the Arctic in summer;
- little seasonal variation of warming at low latitudes or over the southern oceans;
- a reduction in diurnal temperature range over land in most seasons and most regions;
- an increase in anomalously high temperature events and a decrease in anomalously low temperatures;
- an enhanced global mean-hydrological cycle;
- increased precipitation at high latitudes in winter;
- probable increases in intense precipitation events in many regions.

The development of detailed regional scenarios of climate change for use in impact assessment is discussed in more detail in Chapter 6.

CHANGES IN SEA LEVEL

One effect of global warming is to increase sea level, both through thermal expansion of sea water and through the melting of ice on land. Observations from tide gauges indicate that mean global sea level has risen by about 10 to 25 centimetres over the last 100 years, and it appears that this rise is related to the rise in global mean temperature recorded over the same period (see Figure 2.2). Model projections of future global mean sea-level change, based on the temperature change projections given in Figure 2.5, are for a rise of between 13 and 94 centimetres by 2100, with a central estimate of 49 centimetres.[32] Since vertical land movements can also alter the level of the sea relative to that of the land, regional trends in sea level may counteract or reinforce the projected global rise in sea level. However, these projections are clearly of great relevance for vulnerable, low-lying coastal regions and small islands. Furthermore, superimposed on a rising sea level is the possibility of changes in tides, waves and storm surges due to regional climate changes, which could have devastating local consequences. Future projections of these events are still highly uncertain.

Chapter 3

CLIMATE IMPACT ASSESSMENT:

Developing the method

The past 30 years have seen encouraging developments in the method of climate impact assessment. With the benefit of hindsight it seems that our efforts up to the mid 1970s focused mainly on the (one-way) *impact* of climate on human activity. This has recently been replaced and improved upon by a greater emphasis on the *interaction* between climate and human activity.

THE INITIAL EMPHASIS ON IMPACTS

The impact approach is based upon the assumption of direct cause and effect where a climatic event (for example, a short-term variation of temperature) operating on a given exposure unit (for example, a human activity) may have an impact or effect (see Figure 3.1a).[1] This type of analysis may be appropriate for studies of single organisms and the effect of weather, such as high-temperature stress on plants that may cause them to wilt. But larger-scale studies which involve communities of organisms need to consider the processes by which plants may affect weather and climate (for instance, through transpiration). Moreover, many individual organisms are able to adapt to some changes of climate (for example, migration of animals in the face of drought) so that what might be construed as a one-way process of impact is, in reality, a more complex interaction.

Some of the earliest studies of the greenhouse effect and atmospheric ozone depletion were characterized by the impact approach, sometimes using simple regression models to seek a statistical relationship between a climatic event (such as temperature increase) and an inferred impact (such as crop yield). This was true of the Climate Impact Assessment Program funded in the mid 1970s by the US Department of Transportation to evaluate the effect of supersonic aircraft on levels of ozone in the atmosphere.[2] It also characterized a major study by the US National Defense University (NDU) of the possible effects of climate change

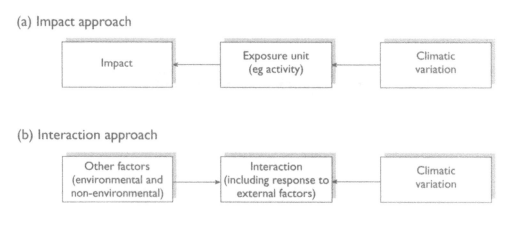

Source: Adapted from Parry and Carter, 1988[3] after Kates, 1985.[4]

Figure 3.1 *Schema of simple* (a) *impact and* (b) *interaction approaches in climate impact assessment*

on agricultural production in the US and the USSR which, it was argued, could affect the balance of power during the Cold War.[5] By comparison with present-day studies, these gave little attention to understanding the processes linking climate, plants and animals. In reality, of course, so many intervening factors operate that it is both misleading and quite impossible to treat these three study elements (climatic variation–exposure unit–impact) in isolation from their environmental and societal milieu, and very few studies have followed this method with success.

THE MORE RECENT EMPHASIS ON INTERACTIONS

More recently, attention has been focused on seeking a better understanding of the interactions between climate and human activity by assuming that a climatic event is merely one of many processes (both societal and environmental in origin) which may affect the exposure unit (Figure 3.1b). To illustrate, it may be argued that the effect of the 1930s drought on the Canadian prairies was substantially increased by the depression of farm prices and the desperate economic straits that had been reached by many Prairie farmers before the beginning of the drought.[6] In addition, widespread ploughing of soils prone to wind erosion increased the amount of windblown dust which choked crops and reduced yields. Economics, weather and farming technology thus interacted to create a severe economic and social impact that was perhaps preconditioned by the Depression but triggered by drought. Likewise, the effects of any climate change in the future will be influenced by concurrent economic and social conditions and the extent to which these create a resiliency or vulnerability to impact from climate change.

This *interaction approach* was first adopted in an explicit form by the project on Drought and Man of the International Federation of Institutes for Advanced Study (IFIAS).[7] Here the emphasis was on comprehending the syndrome of political and economic, as well as meteorological, events which resulted in the severe effects of drought in the Sahel in the mid 1970s. The same approach broadly characterizes many current studies of the effects of greenhouse gas-induced warming (see below).

ORDERS OF INTERACTION

Interaction models achieved greater realism by considering the 'cascade' of interactions that can occur from:

- the first-order biophysical level;
- through second-order levels characterized by units of enterprise (for example, farms and corporations);
- to third-order interactions at the regional and national level (see Figure 3.2).[8]

In the mid 1980s a project based at the International Institute for Applied Systems Analysis (IIASA) worked with hierarchies of models in a number of case studies on the effect of climate change on agriculture.[9] These hierarchies linked models which simulated the response of crops to weather (and which produced estimates of yield) to economic models of the regional economy. In the study based in the Canadian province of Saskatchewan, for example, effects of changes in rainfall were estimated as changes in wheat yield, in farm income and in province-level employment.[10]

The interaction approach also introduced adaptation into climate impact assessment. The IIASA project considered two types of adaptation: adjustments at the enterprise level (which at the farm level might include changes in crops or irrigation) and policy responses at the subnational, national and international level. A schema of the approach is given in Figure 3.3.[11]

THE SEARCH FOR A MORE INTEGRATED APPROACH

Additional complexity can be introduced by studying interactions of the same order, both *within* individual sectors (such as between different farming systems) and *between* different sectors (such as between concurrent effects of climate change on agriculture, forestry and water resources), and the feedback effects operating between them. In theory, an integrated assessment could be performed at any level (biophysical, enterprise or regional) and for any sector, and for any combination of these. Integrated impact assessment, both at

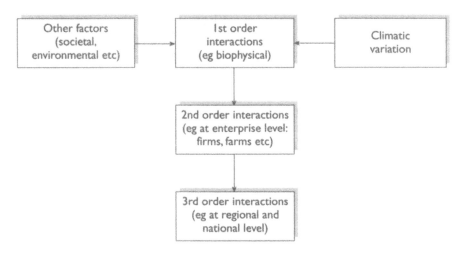

Source: Adapted from Parry and Carter, 1988.[12]

Figure 3.2 *Schema of interaction approach, with ordered interactions*

global and regional levels, is now the subject of considerable attention and is discussed in Chapters 5 and 9.

A major shortcoming of most climate impact assessments to date has been their superficial treatment of adaptation, due in part to its complexity and in part to the lack of a suitable methodological framework. Methods of adaptation assessment are considered in detail in Chapter 9. Impact and adaptation assessment, with broadly equal emphasis, comprises the IPCC approach.

A SEVEN STEP FRAMEWORK FOR ASSESSMENT

The IPCC framework for conducting a climate impact and adaptation assessment is shown in Figure 1.1. It consists of seven main steps which encapsulate the key elements of the approaches described above. The first five steps can be regarded as common to most assessments. Steps 6 and 7 are included in fewer studies. The steps are consecutive (open arrows in Figure 1.1), but the framework also allows for redefining and repeating some steps (thin bold arrows). At each step, a range of study methods is available. These are described and evaluated in the following sections. For reasons of brevity, however, only the essence of each method is introduced, together with references to sources of further information.

Each of the seven general steps includes more detailed procedures, sometimes arranged in a comparable multistep framework. For example, Chapter 9 describes seven equivalent steps in evaluating adaptation strategies. Those steps fit directly into Step 7 of the overall assessment frame-

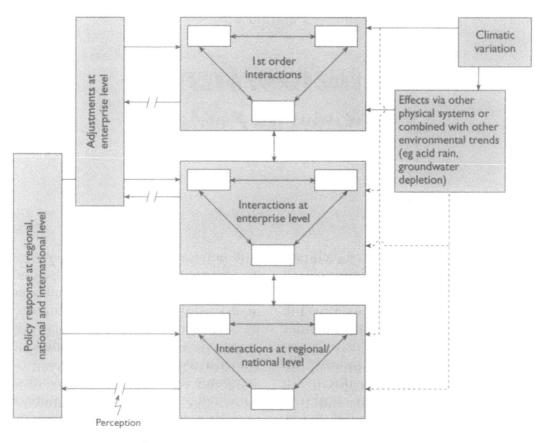

Source: Parry and Carter, 1988.[13]

Figure 3.3 *Schema of IIASA/UNEP project's approach: an interactive approach to climate impact assessment with ordered interactions, interactions at each level and some social and physical feedbacks*

work, but they also parallel all the general assessment steps, because the information required for evaluating adaptation is derived from, and depends on, many of the other steps, such as sensitivity analysis, impact assessment and reliance on specific models.

27

Chapter 4

THE FIRST STEP:

Defining the Problem

A first step in undertaking a climate impact assessment is to define clearly the nature of the problem to be investigated. In order to achieve this it is useful to be clear about: the intended user of the assessment, its purpose, the exposure unit(s) of interest, the spatial and temporal scope of the assessment (its study area and time frame), the data needs, and the wider context of the assessment. We shall take these aspects of problem definition in turn.

THE USES OF CLIMATE IMPACT ASSESSMENTS

Most national impact studies, for example those assessing the effects of greenhouse gas-induced climate change as part of the UN Framework Convention on Climate Change (UNFCCC), will ultimately be used by policy-makers who are unlikely to be expert in the field, though they may be advised by experts. Such assessments thus need to address both an expert and general audience, and they seek to achieve this through reports comprising both technical documents as well as policy-makers' summaries. Assessments covering specific regions or sectors are more likely to be aimed at regional planners who can be expected to have more specialist knowledge. In each case the nature of the assessment, particularly its remit, comprehensiveness and accuracy, should meet the needs of its intended audience. A shortfall in one or more of these may lead to the needs being inadequately addressed, while excess may imply unnecessary expenditure of cost or time.

In general we can discern four specific objectives in climate impact assessment: the detection of causes; the measurement of effects; the identification of vulnerability; and the analysis of adjustments and responses. These are outlined below but are also considered in more detail in subsequent chapters.

The detection of causes

While the assumption in climate impact assessment is that changes or variations in climate or weather are the causes of impact, it is almost certain – under most foreseeable circumstances – that the causes are multiple and interactive. Some may be more direct than others, and they are likely to be both climate- and non-climate related. To be useful, an assessment therefore needs to distinguish between, and preferably quantify, the more and the less important types of causes and to rank in importance the different roles that they play.

The most simple distinction is between direct and indirect causes, although this depends greatly on the purpose of the assessment and thus the viewpoint of the assessor. *Direct or proximate causes* are frequently those events which are proximate in space and time to the effect: they are those which may be considered to have 'triggered' the effect (for example, low night temperatures leading to frost-damaged crops). *Indirect or approximate causes* are those factors which are approximate in time or space to the effect. Approximate factors in time can be those which have preconditioned an especial sensitivity (or vulnerability) to the operation of a proximate factor (for example, the effects of poverty in increasing vulnerability to drought in some developing countries).

Measures of impact

While different measures of impact will be appropriate for different types of assessment, and will also vary according to available data, it is useful to identify at an early stage which measures (given available data, etc) are most suited to the user.

In biophysical assessments, such as the effects on animal and plant growth, there are three broad types of effect:

- the level of productivity of the system (the abundance of a population, or the yield level of a crop plant);
- the range of existence of a population or community (the region inhabitated, the zone of cultivation, etc); and
- the quality of productivity (healthiness of a population, quality of crop yield, etc).

In social and economic assessments the types of effect are less easy to classify, since they are more dependent on the existence of appropriate means of measurement. Quantitative measurements include costs, benefits, levels of income, fertility, morbidity and mortality. It is most important that a minimum set of measurements is established, before the assessment begins, that will meet the needs of the end-user. These are considered in more detail in Chapter 5.

Table 4.1 *The sensitivity to weather of UK industry (£m at 1985 prices)*

	Size	Sensitivity
Agriculture	13,310	4000
Energy	42,900	4250
Minerals, metals manufacturing	42,164	1250
Engineering	76,330	2250
Consumer manufacturing	85,268	6000
Construction	33,946	5000
Retail and distribution	136,641	8000
Transport and communication	14,796	3000
Professional services	150,084	3000
Other services (including TV)	35,157	1500

Note: Left column indicates total value of turnover in each sector. Right column is an estimate of interannual variation in value of turnover due to weather.

Source: Parry and Read, 1988.[1]

Identification of vulnerability

An effective means of identifying those sectors or activities for specific attention in a climate impact assessment is to analyse, first, relative sensitivity to weather on a yearly or seasonal basis. Those activities most sensitive to weather may, a priori, be most vulnerable to climate change.

An illustration is given in Table 4.1, which indicates the varying sensitivity of different industrial sectors in the United Kingdom to weather by assuming that interannual variability in value of turnover is an adequate measure.[2] On this basis and expressed as a proportion of total turnover, agriculture and transport appear the most weather sensitive and thus might warrant more detailed study of the effects of climate change.

Analysis of response

Impact experiments usually evaluate the effects of climate change on an exposure unit in the absence of any responses which might modify these effects. However, a very wide range of responses can be deployed to reduce levels of impact. Two broad types of response can be identified: adaptation and mitigation (see Figure 4.1).[3]

Adaptation is response to the effects of climate change. It may be autonomous (such as a biophysical adaptation of an organism to an altered environment or intuitive reaction by an individual to changed circumstances) or it may be fostered (such as that driven by policy). Many policies of adaptation to climate change make good sense in any case, since present-day climatic variability (in the form of extreme climatic events such as droughts and floods) already causes significant damage in different

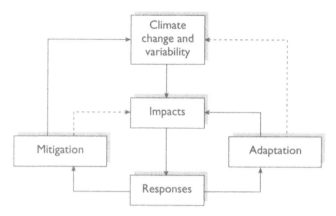

Note: Solid arrows indicate direct effects or feedbacks; dotted arrows depict secondary or indirect effects.
Source: Adapted from Smit, 1993.[4]

Figure 4.1 *Mitigation and adaptation*

parts of the world. Adaptation to these events can thus help to reduce damage in the short term, regardless of any longer-term changes in climate.

Mitigation or 'limitation' attempts to deal with the causes of climate change. It achieves this through actions that prevent or retard the increase of atmospheric greenhouse gas (GHG) concentrations, by limiting current and future emissions from sources of GHGs (for instance, fossil fuel combustion, intensive agriculture) and enhancing potential sinks for GHGs (such as forests and oceans). In recent years there has been a heavy focus on mitigation as a major strategy for coping with the greenhouse problem. However, it seems likely that realistic policies of mitigation will be unable to prevent climate changes, and that alternative adaptive measures are also needed.

While mitigation and adaptation are complementary responses, since both are needed, evaluating mitigation policies is outside the scope of this book. For more information on this topic, the reader is directed to the *Second Assessment Report* of IPCC Working Group II.[5] The evaluation of adaptive responses is considered in Chapter 9.

DEFINING THE TERMS OF REFERENCE

Exposure unit to be studied

The exposure unit to be assessed is likely to determine, to a large degree, the type of researchers who will conduct the assessment, the methods that can be employed and the data required. The choice of exposure unit should reflect the goal of the assessment and the region, group or activity at risk. Studies can focus on a single sector of activity (such as agriculture, forestry, energy production or water resources), or several sectors separately, or several sectors interactively. Alternatively, the exposure unit may be non-sectoral in character (for example, an ecosystem, a distinct regional unit such as an island, or a specific population cohort).

The study area

Selecting a study area is likely to be guided by the goals of the study and by the constraints on available data. Options include:

- administrative units (eg, district, town, province, nation), for which most economic and social data are available and at which level most policy decisions are made;
- geographical units (eg, river catchment, plain, mountain range, lake region), which are useful integrating units for considering multisectoral impacts of climate change;
- ecological zones (eg, moorland, savannah, forest, wetland), which are often selected for considering issues of conservation or land resource evaluation;
- climatic zones (eg, desert, monsoon zone, rain shadow area), which are sometimes selected because of the unique features and activities associated with the climatic regime;
- sensitive regions (eg, ecotones, tree lines, coastal zones, ecological niches, marginal

communities), where changes in climate are likely to be felt first and with the greatest effect;
- representative units, which may be chosen according to any of the above criteria, but are selected to be representative of that regional type and thus amenable to generalization – for instance, a single river catchment may serve as a useful integrating unit in order to consider the impacts of climate on water resources, agriculture, forestry, fisheries, recreation, natural vegetation, soil erosion and hydroelectric power generation; information from this type of study may then be applicable to other similar catchments in a region.

The time frame of the assessment

Selecting a time horizon for study will also be governed, in the main, by the goals of the assessment. For example, in studies of industrial impacts, the planning horizons may be five to ten years, investigations of tree growth may require a 100-year perspective, while considerations of nuclear waste disposal must accommodate time spans of well over 1000 years. However, as the time horizon increases, so the ability to project future trends declines rapidly. Many climate projections rely on general circulation models, which are subject to uncertainties over all projection periods. There is, therefore, an important

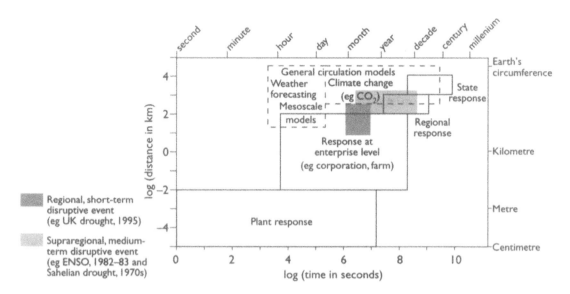

Note: Dotted lines = projections. Solid lines = responses. Shaded areas represent examples of two different scales of disruptive climatic events: a country-level drought (such as in the UK in 1995) and supraregional effects (such as ENSO-related floods and droughts or the 1970s Sahelian drought).

Source: Modified from Parry and Carter, 1984.[6]

Figure 4.2 *The mismatch in time and space scales between* (1) *the present-day ability to project climate changes and* (2) *the likely effects of such changes*

mismatch between our ability to project climate change and the needs of impact assessors. A depiction of this is given in Figure 4.2. Most plant, animal and human responses to weather and climate occur over hours, days and (at most) one or two years, and at the local level. Even relatively large-scale effects such as a regional drought (for example, in the Sahel in the 1970s or the Horn of Africa in the 1980s) occur over a few years and at a regional level; and this is likely to be the scale at which adaptation will be most effective. This mismatch between our ability to forecast and our ability to respond will remain as long as general circulation models (GCMs) have dif-

ficulty in simulating climate change at the regional level. A major effort is now in place to enable 'down-scaling' of results from GCMs to the regional and local level over short timescales.[7]

Data needs

The availability of data limits many impact studies. The collection of new data is an important element of some studies, but most rely on existing sources. Thus, before embarking on a detailed assessment, it is important to identify the main features of the data requirements, namely:

- types of data required;
- time period, spatial coverage and resolution;
- sources and format of the data;
- quantity and quality of the data;
- availability, cost and delivery time of the data;
- licensing and copyright restrictions on data distribution.

There is, of course, a close interdependency between identifying data needs and selecting methods of analysis. In practice, the two procedures operate simultaneously but they are treated consecutively in the IPCC approach for ease of presentation.

Wider context of the work

Although the goals of the research may be quite specific, it is still important to place the study in context, with respect to:

- similar or parallel studies that have been completed or are in progress;
- the political, economic and social system of the study region;
- other social, economic and

environmental changes occurring in the study region;
- issues of scale, where studies conducted at one scale should recognize and take advantage of related information or studies at larger or smaller scales;
- multiple effects of changes in other sectors, in markets or in population;
- the study's policy context.

Considering these aspects will enable the study to be more useful to policy-makers, whose tasks will almost certainly include the need to weigh priorities for public spending in the context of demands from issues other than those relating to climate change.

CONCLUSIONS FOR STEP 1

Most assessments are useful only if they address a specific need. Climate impact assessments are no exception, and those conducting them must be clear about their audience, the issues to be assessed, their areal extent and time frame, and their data requirements.

Chapter 5

THE SECOND STEP:

Selecting the Method

THE CHOICE OF OVERALL APPROACH: TOP-DOWN OR BOTTOM-UP?

The approach traditionally adopted in climate impact assessment is one which assumes certain changes of climate and then evaluates the impacts implied by these. The question posed is: 'for a given climate change, what will be the effect?' We have termed this the direct approach and it has also come to be called the top-down approach.[1] A depiction of it is given in Figure 5.1.[2] The approach is broadly characteristic of recent case studies of climate change impact, such as the US Environmental Protection Agency's series of studies on the global implications of climate change conducted in 1990–93.[3] Here a common set of GCM-based 2 x CO_2 scenarios was adopted in each of five studies on global food supply, sea-level rise, river basins, human health and forests as well as in a regionally integrated impact study in Egypt (see Chapters 8 and 9 for a description of some of these).

More recently, however, the adjoint or bottom-up approach has been developed and implemented. This was introduced at one of three international conferences on climate change held at Villach, Austria, in the 1980s[4] and was subsequently tested in a series of case studies on impacts on agriculture.[5] In the adjoint approach the emphasis is on detecting the sensitivity or vulnerability of the exposure unit to climate change. It is illustrated in Figure 5.2. There are essentially two questions: 'to what aspects of climate change is the exposure unit especially sensitive?' and: 'what changes in these aspects (their magnitude, rate of change, etc) are required to perturb the exposure unit significantly?'

The direct and adjoint approaches are distinguished by major differences in the emphasis on climatic scenarios. The direct approach is largely scenario-driven, while the adjoint approach focuses on the implied effect. The latter is likely to be most appro-

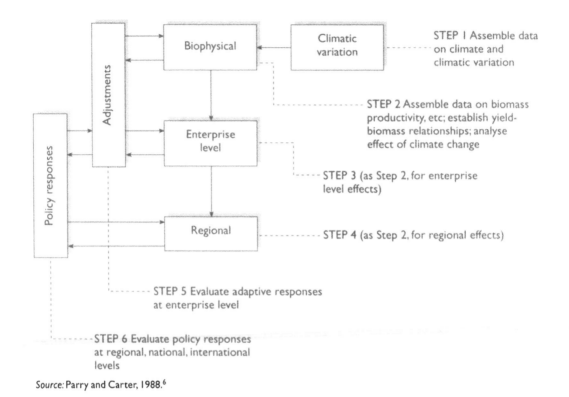

Source: Parry and Carter, 1988.[6]

Figure 5.1 *Direct (or top-down) methods of climate impact assessment*

priate for determining mitigation targets that avoid exceeding damaging levels of GHG concentrations. This is the focal point of Article 3 of the UNFCCC, which requires signatories to avoid GHG concentrations from reaching levels that might threaten natural ecosystems, global food supply or sustainable development.[7]

A combination of the two approaches can effectively be adopted with the following sequence of activities as illustrated in Figure 5.3: firstly, establish the vulnerable exposure units for study (such as agriculture, and water supply); next, identify the types of climate change most likely to affect these; and then, as far as possible, characterize standard sets of GCM-derived climatic scenarios in impact-relevant terms.

METHODS OF ASSESSMENT

Four general analytical methods of climate impact assessment can be identified: experimentation, modelling, empirical analogue studies and expert judgement.

Experimentation

In the physical sciences, a standard method of testing hypotheses, or of

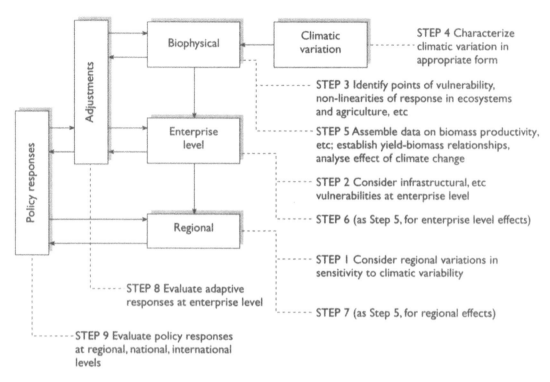

Source: Parry and Carter, 1988.[8]

Figure 5.2 *Adjoint (or bottom-up) methods of climate impact assessment*

evaluating processes of cause and effect, is through direct experimentation. In the context of climate impact assessment, however, experimentation has only a limited application. Clearly, it is not possible physically to simulate large-scale systems such as the global climate, nor is it feasible to conduct controlled experiments to observe interactions involving climate and human-related activities. Only where the scale of impact is manageable, the exposure unit measurable, and the environment controllable can experiments be usefully conducted.

Up to now most attention in this area has been on observing the behaviour of plant species under controlled conditions of climate and atmospheric composition.[9] In the field such experiments have mainly comprised gas-enrichment studies, employing gas releases in the open air, or in open or closed chambers including greenhouses. The former experiments are more realistic but are also more expensive and less amenable to control. The chamber experiments allow for climatic as well as gas control, but the chambers may introduce a new set of limiting conditions which would not occur in reality. The greatest level of

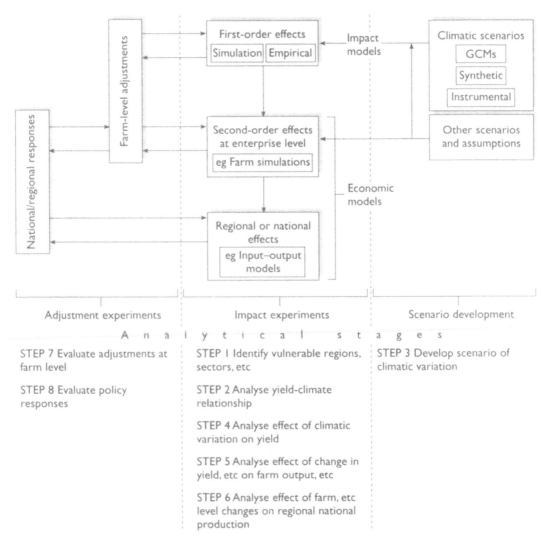

Note: This example is taken from the IIASA/UNEP project on agriculture.

Source: Adapted from Parry and Carter, 1988.[10]

Figure 5.3 *The combination of direct and adjoint approaches in climate impact assessment*

control is achieved in the laboratory, where processes can be studied in more detail and more sophisticated analyses can be employed.

The primary gases studied have been carbon dioxide, sulphur dioxide and ozone, all of which are expected to play an interactive role with climate in future plant growth and productivity. Both temperature and water relations have also been regulated, to simulate possible future climatic conditions. To date, there have been experiments with agricultural plants (both annual and perennial crops), crop pests and diseases (often in conjunction with host plants), trees (usually saplings, but also some mature species), and natural vegetation species and communities (where aspects of competition can be studied). Controlled experiments have also been reported on freshwater ecosystems (to study effects on water quality and the food chain) and soils (examining decomposition rates, nutrient leaching and microbial activity).

There are other sectors in which experimentation may yield useful information for assessing impacts of climatic change. For instance, building materials and design are continually being refined and tested to account for environmental influences and for energy-saving. Information from these tests may provide clues as to the performance of such materials, assuming they are widely employed in the future, under altered climatic conditions. The information obtained from experiments, while useful in its own right, is also invaluable for calibrating models which are to be used in projecting impacts of climate change (see below).

Modelling

One of the major goals of climate impact assessment, especially concerning aspects of future climate change, is the prediction of future impacts. A growing number of model projections have become available on how global climate may change in the future as a result of increases in GHG concentrations.[11] These results, along with scientific and public concerns about their possible implications, have mobilized policy-makers to demand qualitative assessments of the likely impacts within the time horizons and regional constraints of their jurisdiction. Thus, a main focus of much recent work has been on impact projections, using an array of mathematical models to extrapolate into the future. In order to distinguish them from 'climate models', which are used to project future climate, the term 'impact model' has now received wide currency.

At the start of any climate impact assessment, researchers are commonly confronted with an important choice with regard to impact models – either to adopt existing models or to develop new models. Bearing in mind that most assessments have severe time and resource constraints, the most sensible strategy for model selection is, first, to conduct a rigorous survey of existing models that apply to the issue being investigated. This exercise is best conducted by experienced modellers, but some information for non-specialists can also be provided by international organizations, who can advise on suitable models or even supply them directly. Examples of these can be found in the following sections.

The second important step is to examine a model's data needs.

39

Without suitable input data, even the most informative of models cannot be used. If there are suitable data, the models can be tested according to the procedures described in Chapter 5. If input data are not available, or inadequate, then for some applications it may be necessary or desirable to collect the appropriate information.

Finally, if suitable models cannot be identified, then it may become necessary to develop new ones. In some regions with appropriate data it may be possible, in quite a short time, to construct simple statistically based models which are robust enough to apply to climate change problems. This has often been the practice in many less developed countries, where access to more sophisticated models is sometimes limited, and the development of such models may be constrained by poor data quality and lack of modelling expertise. Even in developed countries, however, in the context of an impact assessment study, construction of more complex models from first principles is likely to be too time and resource intensive and is rarely undertaken. It is more common for model development to involve refinements of existing models which take account of altered conditions under a changing climate. For example, many crop-growth models developed for yield prediction under present-day conditions have been modified for climate impact studies to account for the effects of increasing CO_2 on carbon uptake and water use (assumed constant in conventional applications).

Some of the specific procedures for projecting future impacts are described in Chapters 7 and 8. Here, the major classes of predictive models and approaches are explored. It is convenient, in categorizing impact models, to follow the hierarchical structure of interactions that was introduced in Chapter 3. Direct climatic effects are usually assessed using biophysical models, while indirect or secondary effects are generally assessed using a range of biophysical, economic and qualitative models. Finally, attempts have also been made at comprehensive assessments using integrated systems models. We shall consider these in turn.

Biophysical models
Biophysical models are used to evaluate the physical interactions between climate and an exposure unit. There are two main types: empirical–statistical models and process-based models. The use of these in evaluating future impacts is probably best documented for the agricultural sector,[12] the hydrological aspects of water resources[13] and ecosystems,[14] but the principles can readily be extended to other sectors.

Empirical–statistical models are based on the statistical relationships between climate and the exposure unit. They range from simple indices of suitability or potential (for instance, identifying the temperature thresholds defining the ice-free period on important shipping routes), through univariate regression models used for prediction (for example, using air temperature to predict energy demand) to complex multivariate models, which attempt to provide a statistical explanation of observed phenomena by accounting for the most important factors (such as, predicting crop yields on the basis of temperature, rainfall, sowing date and fertilizer application). Such models are usually developed on the basis of present-day climatic variations. Thus, one of their major weaknesses in considering future climate change is their limited ability to predict the effects of climatic events that lie outside the range of present-day variability. They may also be criticized for being based on statistical relationships between factors rather than on an understanding of the important causal mechanisms. However, where models are founded on a good knowledge of the determining processes, and where there are good grounds for extrapolation, they can still be useful predictive tools in climate impact assessment.

Empirical–statistical models are often simple to apply and less demanding of input data than process-based models. For example, the effect of drought on wheat yield on the US Great Plains has been calculated for each of 53 crop-reporting districts, using a regression equation which expressed the relationship between actual yield and the weather experienced in each district in the 1930s.[15] Assuming a recurrence of 1930s conditions with 1975 technology, expected yields were thus mapped, relatively

quickly, for a major wheat-producing region of the world (see Figure 5.4). The obvious drawback to this approach is the assumption of fixed technology and the use of the 1930s drought as an analogue of GHG-induced climate change. The latter, however, has the advantage of being based on a credible scenario (the climate conditions actually occurred in the past), while the former can be improved upon by allowing for developments in agricultural technology. More recent studies of climate change in the Great Plains have combined process-based impact models with the 1930s analogue climate scenario (for a description of this, see Chapter 9 and Box 9.3).

A particular type of empirically derived coefficient, especially useful for broad-scale mapping of areas of potential impact, is the *climatic index*. This is a derived variable that is defined either by manipulating values of a meteorological variable into a different form or by combining variables into a composite term. Perhaps the most common climatic index is a sum of temperature over time, often expressed in degree-days. This computes the temperature accumulated above or below a temperature threshold. The threshold may relate to the onset of plant growth (growing degree-days), or to the requirement for comfortable levels of space heating (heating degree-days), or cooling for air conditioning (cooling degree-days). Figure 5.5 illustrates changes in heating degree-days averaged over

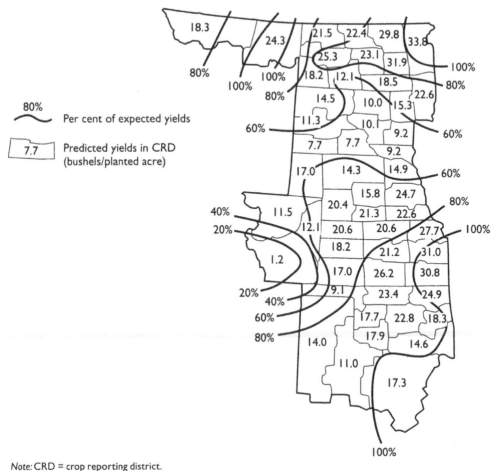

Note: CRD = crop reporting district.

Source: Warrick, 1984.[16]

Figure 5.4 *Simulated US Great Plains wheat yields assuming a recurrence of 1936 drought conditions with 1975 technology*

the UK and weighted according to regional energy use under increases in temperature assumed by the UK's Climate Change Impacts Review Group for the years 2000, 2020 and 2050.[17] A conclusion from this is that demand for space heating may decline by 6 per cent by 2000, 11 per cent by 2030 and 16 per cent by 2050 due to GHG-induced climate change. The demand for air conditioning would increase, but by a smaller amount.

Broadscale mapping of the potential impacts of climate change can be achieved through mapping the spatial shift of isolines of climatic indices under different climatic scenarios; this provides a preliminary, low-resolution but rapid means of identifying areas most suitable for study in more detail. An example of this approach is the shift under altered climate of cli-

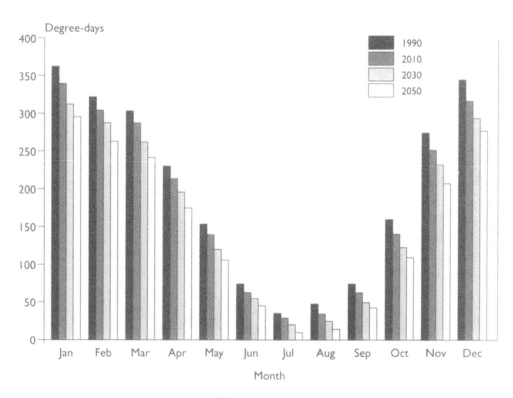

Source: UK Department of the Environment, 1996.[18]

Figure 5.5 *Heating degree-days in the UK as a result of climate change, 1990, 2010, 2030 and 2050*

matic indices that broadly define the range of the anopheles mosquito (see Figure 5.6).[19] While many other factors are involved in explaining the implications of climate change for the distribution of malaria, mapped indices of the altered range of the disease vector are a useful starting point. More detailed process-based models of, amongst other things, insect behaviour, are needed to provide a more comprehensive insight into the potential effects of climate change on this disease and on other risks to human health.

Process-based models make use of established physical laws and theories to express the interactions between climate and an exposure unit. In this sense, they represent processes that can be applied universally to similar systems in different circumstances. For example, there are well-established methods of modelling leaf photosynthesis which apply to a range of plants and environments. Usually some kind of model calibration is required to account for features of the local environment that are not modelled explicitly, and this is gener-

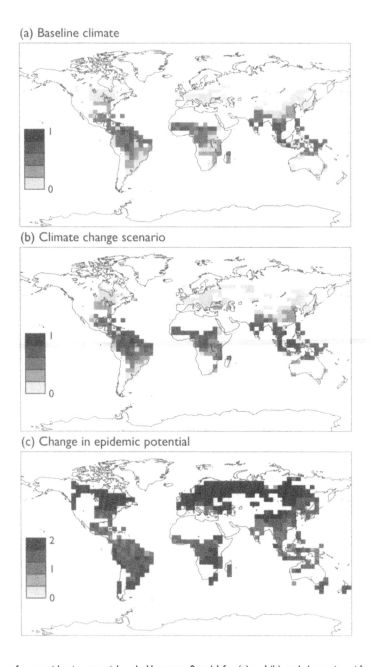

(a) Baseline climate

(b) Climate change scenario

(c) Change in epidemic potential

Note: Numbers refer to epidemic potential, scaled between 0 and 1 for (a) and (b), and change in epidemic potential (c).

Climate change scenario is based on the climate patterns generated by the ECHAMI-A GCM.

Source: Martens, 1997.[20]

Figure 5.6 *Potential malaria risk areas (a) for baseline climate conditions (1931–80), (b) for a 1.2°C temperature increase and (c) changes in average annual epidemic potential, a measure of vectoral capacity, relative to baseline climate (for P falciparum)*

ally based on empirical data. Nevertheless, there are often firmer grounds for conducting predictive studies with these process-based models than with empirical–statistical models. The major problem with most process-based models is that they generally have demanding requirements for input data, both for model testing and for simulating future impacts. This tends to restrict the use of such models to only a few points in geographical space where the relevant data are available. In addition, they are seldom able to predict system responses successfully without considerable efforts to calibrate them for actual conditions. For example, crop yields may be overestimated by process-based yield models because the models fail to account for all of the limitations on crops in the field at farm level, such as weeds and insect pests.

Over the past decade many process-based models have been applied in climate impact assessment, but documentation on the models is often poor or difficult to obtain, and the selection of appropriate models for a particular problem or region can be very difficult. Recently, efforts have been made to organize model intercomparison exercises. These compare the performance of different models based on identical model inputs and assumptions. Examples include comparisons of evapotranspiration models,[21] crop growth models for wheat[22] and potato,[23] and global vegetation–ecosystem models.[24] Such

exercises, combined with efforts to make alternative models available to users within a standard operating environment, offer a ready opportunity for potential users to test and apply different models themselves.

For example, the IBSNAT/ICASA models for the major food staples (including wheat, rice, maize, sorghum, millet, soya) have been used in several international climate impact assessments because they are well documented, relatively easy to use and enable a degree of comparability to be achieved region-to-region and crop-to-crop.[25] As an example of the use of process-based models, Box 5.1 describes the use of the IBSNAT/ICASA models in further detail. Some results obtained from their use are discussed in Chapters 8 and 9.

Economic models
Economic models of many kinds can be employed to evaluate the implications of climate change for local and regional economies. To simplify their classification, it is useful to distinguish between three types of economic models, according to the approach used to construct them, and three scales of economic activity that different model types can represent.

Programming models have an objective function and constraints. The objective function represents the behaviour of the producer (such as profit maximizing or cost minimizing). If the objective function and constraints are linear, the model is known as a linear programming (LP)

BOX 5.1
THE IBSNAT/ICASA CROP MODELS

Background: the IBSNAT/ICASA models employ simplified functions to predict the growth of crops as influenced by the major factors that affect yields, such as genetics, climate (daily solar radiation, maximum and minimum temperatures and precipitation), soils and management. Models are available for wheat, barley, maize, paddy and upland rice, soybean, sorghum and millet, peanut, drybean and potato. The IBSNAT/ICASA models have been tested over a wide range of environments and are not specific to any particular location or soil type. Thus they are suitable for use in international studies in which crop-growing conditions differ greatly. Furthermore, because management practices, such as cultivar, planting date, plant population, row spacing and sowing depth, may be varied in the models, they permit experiments that simulate management adjustments by farmers to climate change.

Description: the models describe daily phenological development and growth in response to environmental factors (soils, weather and management). Modelled processes include phenological development (duration of growth stages), growth of vegetative and reproductive plant parts, extension growth of leaves and stems, senescence of leaves, biomass production and partitioning among plant parts, and root system dynamics. The models include subroutines to simulate the soil and crop water balance and the nitrogen balance.

Development: the primary variable influencing phasic development is temperature. The thermal time for each phase is defined by coefficients that characterize, for example, the duration of the juvenile phase, photoperiod sensitivity and duration of the reproductive phase.

Dry matter production: potential dry matter production is a linear function of intercepted photosynthetically active radiation (PAR). The percentage of incoming PAR intercepted by the canopy is an exponential function of leaf area index. The dry matter allocation is determined by partitioning coefficients which depend on phenological stage and degree of water stress. Final grain yield is the product of plant population, kernels per plant and kernel weight.

Carbon dioxide sensitivity: the IBSNAT/ICASA models have been modified to simulate the changes in photosynthesis and evapotranspiration caused by higher levels of CO_2. Ratios have been calculated between measured daily photosynthesis and evapotranspiration rates for a canopy exposed to a range of high CO_2 values, based on published experimental results.

Input data: the models require daily values for solar radiation, maximum and minimum temperature and precipitation. Soil data for different layers are needed to compute drainage, runoff, evaporation, soil water-holding capacities and rooting preference coefficients. Initial soil water content should also be specified. The input data for all models are formatted in the same way.

Output files: a set of output files, identically formatted for all models, is created

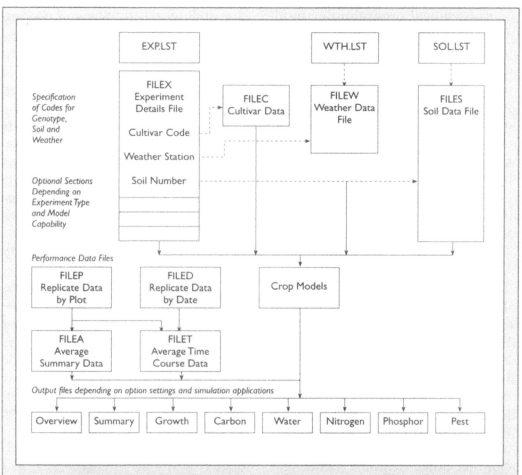

during each simulation run. The first output file saves the information that is displayed on the computer monitor during the simulation. Subsequent output files contain predicted variables related to growth, carbon balance, soil water, plant and soil nitrogen and phosphorous and an optional file on pest risk. A final output file contains a one line summary for each simulation run. Variables included in the summary output file are the main phenological events, yield and yield components (see figure).

model. This is the most common type of programming model used in climate impact assessments. LP models can also incorporate nonlinear relations (for example, technical relations) in a piecewise manner. Programming models can be of the partial equilib-rium type: they can determine production (supply) and demand simultaneously. They are usually calibrated to a set of data in a given year. In this sense they are empirically based. Programming models can be static or dynamic.

Those linear programming models most widely used in climate impact assessment have been designed to predict changes in land use under altered conditions of supply or demand. Studies of the impact of climate change on land use have now been completed for England and Wales[26] and the United States.[27] In both cases the LP models used were ones which optimized the allocation of land under estimated changes in yield (and thus profitability) due to climate change. In the England and Wales study, land was classified into 32 classes of suitability at a resolution of 1 kilometre. The 150,000 kilometre squares in England and Wales were then treated as individual fields within a single 'national farm', each first being allocated to its highest value-use under limits of demand and under current climate, and then being reallocated to different uses as a result of changes in yield due to climate change. Since 'highest value-use' depends not only on yield but also on price (of inputs as well as outputs), the model needs to account for changes in global food prices due to climate-induced changes in global food production as well as yield changes due to climate change over England and Wales alone. A schema of this approach is given in Figure 5.7.

Econometric models consist of supply and/or demand functions which use prices as independent variables as well as a number of technical variables, and usually include time to represent those parts of the economy that undergo steady change. Like programming models, these models also have their parameters numerically quantified, but econometric models differ substantially in their structure from programming models. Conventionally, econometric models do not state any decision rules. However, in the last decade a new set of econometrically specified models has emerged: the so-called dual models. These assume decision rules such as profit maximizing or cost minimizing of producers and utility maximizing or expenditure minimizing of the consumer. In these cases, data fitting is usually done by statistical methods (regression analysis) or a simple calibration procedure. The bulk of econometric models are static (including those that embed a time trend), although among the few examples of dynamic models are the so-called adaptive models.

Input–output (IO) models are developed to study the interdependence of production activities. The outputs of some activities become the inputs for others, and vice versa.[28] These input–output relationships are generally assumed to be constant, which is a weakness of the approach, since reorganization of production or feedback effects (such as between demand and prices) may change the relationships between activities. This is of particular concern when projecting production activities beyond a few years into the future. More recently, dynamic versions of IO models have been developed, but these still lack many of

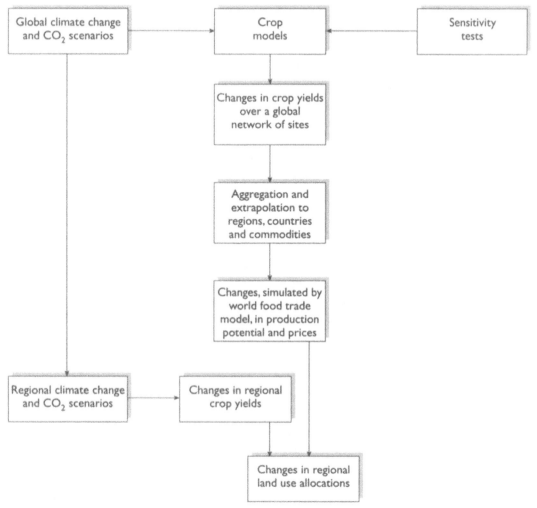

Note: The project considered land use changes in England and Wales.

Source: Parry et al, 1996.[29]

Figure 5.7 *The use of a linear programming model for optimized land allocation under conditions of altered climate*

the dynamic aspects of economic behaviour. Nonetheless, the approach is relatively simple to apply and the data inputs are not demanding. Moreover, these models are already in common usage as planning tools. Examples of their application in climate impact assessment include studies of

possible impacts of climate change on the economy of Saskatchewan[30] and on economic activity in the states of Missouri, Iowa, Nebraska and Kansas (the MINK study) in the US (see Box 9.3 in Chapter 9).

Three scales of economic activity are commonly represented by economic models: firm-level, sector-level and economy-wide. *Firm-level models* depict a single firm or enterprise (that is, a decision unit for production). These are often programming models but are rarely of the econometric type, due to constraints on available information about firms. Typical examples include farm level simulation models, which attempt to mirror the decision processes facing farmers who must choose between different methods of production and allocate adequate resources of cash, machines, buildings and labour to maximize returns.[31] Such models may also require data on productivity, and it is this which constitutes the entry point for potential linkages with the outputs from biophysical models. Model outputs include farm-level estimates, for example, of income, cash flow and resource costs for obtaining selected production plans. These models are sometimes referred to as microsimulation models.

Sector-level models encompass an entire sector or industry. They can be programming models, or of the econometric type, that depict production. For climate change studies these models should be of a partial equilibrium type, to include demand so that price changes are generated as well. It is quite common for such models to consider a firm as representing the average of the entire sector under study. Such models are then similar to firm-level models, but require aggregation and assumptions about average technical relations. Some sec-

tor-level models are also of the IO type and have supply and demand included. These models usually have no or very few links to developments in the rest of the economy.

Economy-wide models, sometimes referred to as macroeconomic models (which are actually a large subset of this class), link changes in one sector to changes in the broader economy, dealing with all economic activities of a spatial entity such as a country, a region within a country or a group of countries. Typical economy-wide models for climate impact assessment include all types of general equilibrium (GE) models and IO models. Most GE models belong to the group of dual econometric models, but they also include programming models. The distinctive feature of GE models is that they compute prices which clear the market in the same way as partial equilibrium models. However, unlike partial equilibrium models, GE models encompass all economic activities of the region. The static form of the GE model is the computable general equilibrium (CGE) model. Some of the studies of climate impacts conducted to date with CGE models have used, as inputs, the results of studies of sectoral impact. For example, the results of an agricultural impact study by Adams et al,[32] together with results from studies on coasts (related to sea-level rise) and electricity demand, were used as inputs to a general equilibrium model of the US economy to assess the wider implications in all sectors of the economy.[33]

There are also dynamic GE models, which can treat the evolution of an economy through time, ensuring at each step that the markets are in equilibrium. For example, a (recursively) dynamic GE model of global food trade, the Basic Linked System (BLS), has been used to study the potential effects of climate change on global food supply, as part of the US Environmental Protection Agency (EPA)-funded international impacts projects.[34] This was based on information on the potential yield changes of major crops taken from IBSNAT/ICASA crop-modelling studies which were conducted at 112 sites in 18 countries. The estimates of climate change-induced yield changes were interpolated from site results across the geographical area of all countries and from the modelled crops, to crops and livestock not modelled (such as cassava and oil palm) – see Figure 5.8. The BLS model used these inputs, together with assumed changes over time (to 2060) in demand (through population growth and per capita consumption resulting from economic growth) and in supply (through technological change leading to increased yields), to estimate changes in food prices that might result from climate change. The results of the research are discussed further in Chapters 8 and 9. Once a global estimate of impact (in this example, on food prices) has been obtained, this information can serve as an exogenous control in regional impact studies. Examples of linked global and regional studies are those for land use in England and Wales described above and for agriculture in Egypt, illustrated in Box 9.1.

Economic models are the only credible tools for deriving meaningful estimates of the likely effects of climate change on measurable economic quantities, such as income, gross domestic product (GDP), employment and savings. However, great care is required in interpreting the results. Specifically, caution must be exercised in using any of the measures of economic activity as indicators of social welfare. Potentially more serious, however, is the failure of most models (exceptions include the models of Cline and Fankhauser)[35] to account for non-market effects of climate change. For example, many inputs to production are directly affected by climate change (such as land and water) but are not contained in most macroeconomic models. Economic models are also widely used to consider the relative cost-effectiveness of mitigation and adaptation options that are proposed to ameliorate the adverse impacts of climate change, along with the associated economic, social and environmental impacts of these options. Some of these points are further addressed below in relation to integrated models.

Integrated systems models
The issue of greenhouse gas-induced climate change now assumes a high profile in national and international

Source: Adapted from Rosenzweig et al, 1993.[36]

Figure 5.8 *Use of a general equilibrium model of the world food trade (the Basic Linked System) in a study of the effects of climate change on food supply and risk of hunger*

policy-making. In order to inform policy, however, it is necessary to identify and address all of the different components of the problem. This has been the motive force behind recent efforts to integrate the causes, impacts, feedbacks and policy implications of the 'greenhouse problem' within a modelling framework. Two main approaches to integration can be identified: the aggregate cost-benefit approach and the regionalized process-based approach.

The *aggregate cost-benefit approach* seeks to estimate the likely monetary costs and benefits of GHG-induced climate change in order to evaluate the possible policy options for mitigating or adapting to climate change. This is a macroeconomic modelling approach (see above) and has been applied to certain aspects of the greenhouse problem for many years. In particular, the methods have been used to compute the development paths of carbon dioxide and other greenhouse gas emissions in the atmosphere (the driving force for climate change). The approach commonly combines a set of economic models with a climate model and a damage assessment model. The economic models provide global projections (sometimes disaggregated into major regional groupings of countries) of supply and demand in commodities that can affect greenhouse gas emissions, on the basis of future world population and economic development. The models use price to determine the relative com-

petitiveness of different technologies of energy production, while accounting for the long-term depletion of fossil fuels, allowing for the development of more efficient technologies, and accommodating likely policies of emissions abatement. The time horizon considered can range from a few decades to several centuries.

Climate models refer to a suite of functions that are needed: firstly, to convert GHG emissions into atmospheric concentrations; secondly, to estimate radiative forcing of the climate; and thirdly, to compute the climate sensitivity of the forcing (global mean temperature response to radiative forcing equivalent to a doubling of CO_2). Such functions usually comprise simplified representations of the gas cycles, empirical methods of determining radiative forcing, and highly simplified equations for computing temperature response.

Damage-assessment models are functions that provide an estimate of the likely impacts (costs) of climate change, usually as a percentage of gross national product (GNP). They commonly provide a global estimate of 'damage' as a function of global mean temperature change. To date, such functions have been chosen subjectively on the basis of expert opinion, or using the few quantitative estimates that are available on the possible sectoral impacts of climate change at the global scale. Great caution must be exercised, however, since simulation outcomes with these models can be very sensitive to assumptions, such as those concerning future discount rates and the estimated damage response. A further major difficulty is assigning a value to intangible non-market goods such as human well-being, a pollution-free environment and biological diversity. Recent examples of models exhibiting this type of three-component framework include DICE,[37] CETA[38] and MERGE.[39] A thorough review of this type of model, collectively referred to as 'policy optimization models', has been conducted by Working Group III of the IPCC,[40] and a list of models is presented in Appendix 1.

The *regionalized process-based approach* attempts to model the sequence of cause and effect processes originating from scenarios of future GHG emissions, through atmospheric GHG concentrations, radiative forcing, global temperature change, regional climate change, possible regional impacts of climate change and the feedbacks from impacts to each of the other components. Regional impacts can be aggregated, where appropriate, to give global impacts which can then be used to evaluate the likely effectiveness of global or regional policies. The approach is derived from the applied natural sciences, especially ecology, agriculture, forestry and hydrology, where climate impact assessment has evolved from site or local impact studies towards large area assessments, using process-based mathematical models in combination with geographical information sys-

tem (GIS) technology. The first attempt at a spatially explicit, integrated representation of the climate change issue, from sources to impacts, was the model IMAGE 1.0,[41] which was subsequently incorporated within an integrated model for Europe, ESCAPE.[42] During the 1990s, several other models of this kind were developed. Box 5.2 illustrates an application of IMAGE 2.0, probably the most sophisticated model yet to have been developed.[43]

In contrast to the aggregate cost-benefit approach, the estimates of biophysical impacts in these models are quantitative and regionally explicit. In addition, the treatment of gas cycling and climate change is usually more sophisticated than in the former approach. The economic impacts of climate change are not yet incorporated, however, and future versions of these models will strengthen their regional economic and global trade components, thus offering a quantitative assessment of the 'damage' described above. Some of these developments are discussed further in a review of these models undertaken by IPCC Working Group III, where they fall into the category of 'policy evaluation models'.[44]

The two types of approach outlined above originate from quite different disciplinary perspectives and were developed for contrasting reasons. However, it is becoming increasingly evident that major refinements of one approach will require significant contributions from the other. Indeed, it appears that the two approaches are rapidly converging towards a common, interdisciplinary method that will become a standard tool in policy analysis. Nevertheless, there are numerous problems associated with integrated systems models, including their complexity, lack of transparency and demanding data requirements for calibration and testing. Moreover, a major concern remains about the ability of these models to represent the uncertainties that propagate through each level of the modelled system.

Empirical analogue studies

Observations of the interactions of climate and society in a region can be of value in anticipating future impacts. The most common method employed involves transferring information from a different time or place to an area of interest to serve as an analogy. Four types of analogy can be identified: historical event analogies; historical trend analogies; regional analogies of present climate; and regional analogies of future climate. Analogues can also be used as climatic scenarios (see Chapter 7).

Historical event analogues
Historical analogues use information from the past as an analogue of possible future conditions. Data collection may be guided by anomalous climatic events in the past record (for instance drought or hot spells) or by the impacts themselves (periods of severe

soil erosion by wind). An example is the use of the 1930s drought as an analogue scenario of a greenhouse-gas-induced climate change in the US Great Plains (see Figure 5.4).

Historical trend analogues
Historical trend analogues use past trends which are unrelated to greenhouse gases as an analogy of GHG-induced change. Long-term temperature increases due to urbanization are one potential source for a warming analogue (as yet seldom considered by impact analysts). Another case is past land subsidence, the impacts of which have been used as an analogue of future sea-level rise associated with global warming. For example, much of the city of Bangkok, Thailand, is only one metre above sea level. In recent decades, increased extraction of water from local aquifers to supply a rapidly growing population has led to severe subsidence, with rates of 100 millimetres per year in central Bangkok and 50 to 100 millimetres per year in the suburbs. The impacts already experienced in Bangkok include flooding due to river overflow, heavy rain and high tides, damage to drainage systems, damage to structures, and saltwater intrusion.[45]

Regional analogues of present climate
Regional analogues of present climate are areas which have a similar present-day climate to the study region and for which the impacts of a given climatic anomaly, such as drought, can be com-

BOX 5.2

AN APPLICATION OF IMAGE 2.0: A GLOBALLY INTEGRATED
SYSTEMS MODEL

Background: IMAGE 2.0 is a global model designed to provide a science-based overview of climate change issues to support the national and international evaluation of policies (Alcamo, 1994; Alcamo et al, 1996).[46]

Model: IMAGE 2.0 consists of three fully linked components – energy–industry, terrestrial environment and atmosphere–ocean (see figure). Dynamic calculations are performed for a 100-year time horizon and the model is embedded in a geographical information system. The energy–industry set of models is used to compute the emissions of greenhouse gases in each region as a function of energy consumption and industrial production. The terrestrial environment component simulates land use and land cover dynamically through time over a 0.5° x 0.5° latitude–longitude grid, employing these changes to determine greenhouse gas emissions from the terrestrial biosphere to the atmosphere. The atmosphere–ocean set of models computes the build-up of greenhouse gases in the atmosphere and the resulting change in climate. Emissions from the energy–industry and terrestrial environment components are combined and used to determine the uptake of carbon by the oceans and the atmospheric gas and aerosol composition. The climatic response to atmospheric forcing is determined with an atmospheric energy–balance model, which is used in conjunction with information from GCMs to provide regional climate change scenarios.

Application: this example illustrates how the model determines feedback processes in the response of the terrestrial carbon cycle to climate change.

Methods: the terrestrial environment component of IMAGE 2.0 was used to compute the carbon fluxes between the terrestrial biosphere and the atmosphere. The model can simulate the effects of feedback processes occurring under increased atmospheric CO_2 concentrations and a changing climate: the enhancement of plant growth (CO_2 fertilization) and increased water use efficiency; temperature responses of plant photosynthesis and respiration; temperature and soil water responses of decomposition processes; and climate-induced changes in vegetation and agricultural patterns and consequent changes in land cover. A unique feature of the model is its ability to relate changes in land cover to the demand for agricultural land. This component is driven by regional population and economic activity. The agricultural demands are combined with regional potential crop productivity and distribution to determine the amount of agricultural land required. If this exceeds the current amount, simple rules are applied to determine the expansion of agricultural land into areas currently under other land cover types (for example, using the nearest areas with the highest potential productivity first).

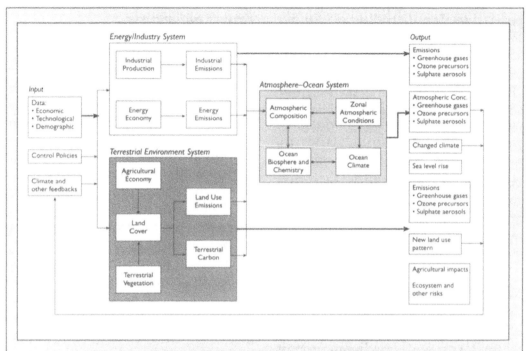

Schematic diagram of IMAGE 2.0 model

Scenarios: the projection horizon is 1970 to 2050. The IPCC best estimate scenario (IS92a) is used to define the socio-economic projections: a world population increase of 93 per cent and GNP increase of 134 per cent by 2050. The climatic scenario is based on the Geophysical Fluid Dynamics Laboratory (GFDL) $2 \times CO_2$ equilibrium experiment (Manabe and Wetherald, 1987),[47] assumed to be concurrent with an equivalent CO_2 concentration of 686 ppm by 2050 (570 ppm for CO_2 alone) relative to 1970.

Impacts: changes in climate and in water use efficiency induce shifts in vegetation patterns relative to 1970. CO_2 fertilization decreases net carbon emissions to the atmosphere while changed decomposition rates increase emissions, though regionally there are large differences. Changes in the global balance between photosynthesis and respiration make little net difference. Neglecting land use changes, the terrestrial biosphere acts as a net carbon sink (negative feedback) relative to the current situation. However, with increasing population, the demand for new agricultural land is large, and land cover changes with associated carbon emissions are likely to counteract completely the negative feedbacks described above.

Source: Vloedbeld and Leemans, 1993.[48]

pared directly with the impacts under more normal climatic conditions in the study region. The advantage of this method is that the comparison is synchronized in time, so that the wider socio-economic context in which the impacts occur is equivalent. However, for the approach to be valid, the impacts of climate should be similar in both regions, implying that their environmental conditions (such as soils and topography), their level of development and their respective economic systems also exhibit similarities.

Regional analogues of future climate

Regional analogues of future climate work on the same principle as analogues for present-day climate, except that here the analyst attempts to identify regions which have a climate that is similar to that projected for the study region in the future. This principle has proved valuable in extending the range of applicability of some impact models. For example, a model of grass growth in Iceland has been tested for species currently found in northern Britain, which is an analogue region for Iceland under a climate some 4°C warmer than present.[49] Analogues of this kind can help to identify the kinds of adaptation that may prove most appropriate. Figure 5.9 shows regional analogues for one GCM-derived 2 x CO_2 scenario – in this case estimated as part of the IIASA / UNEP study in the mid 1980s.

Expert judgement

A useful method of rapidly assessing the state of knowledge concerning the effects of climate on given exposure units is to solicit the judgement and opinions of experts in the field. Of course, expert judgement plays an important role in each of the other analytical methods described above. On its own, however, the method is widely adopted by government departments for producing position papers on issues requiring policy responses. In circumstances where there may be insufficient time to undertake a full research study, literature is reviewed, comparable studies identified, and experience and judgement are used in applying all available information to the current problem.

The use of expert judgement can also be formalized into a quantitative assessment method, by classifying and then aggregating the responses of different experts to a range of questions requiring evaluation. This method was employed in the National Defense University's (NDU) study of 'Climate Change to the Year 2000', which solicited probability judgements from experts about climate change and its possible impacts.[50] The pitfalls of this type of analysis have been examined in detail in the context of the NDU study.[51] They include problems of questionnaire design and delivery, selection of representative samples of experts, and the analysis of experts' responses.

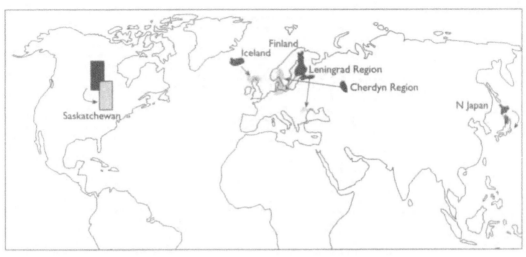

Source: Parry and Carter, 1988.[52]

Figure 5.9 *Present-day analogue regions of the Goddard Institute for Space Studies (GISS) 2 x CO$_2$ climate (Hansen et al, 1983)[53] for Saskatchewan (Canada), Iceland, Finland, the Leningrad and Cherdyn regions of Russia and the Hokkaido and Tohoku regions of Japan*

More recently, decision support systems that combine dynamic simulation with expert judgement have emerged as promising tools for policy analysis. Here, subjective probability analysis is required where process-based or empirical models are lacking. An example of this approach, employing a Bayesian network analysis of expert knowledge to evaluate the uncertainties of climate change impacts on aquatic systems, has been presented for Finland.[54] Participatory assessment, involving a range of stakeholders at different stages of a regional impact study, is another approach which has been tested in the Mackenzie Basin study in Canada.[55]

CONCLUSIONS FOR STEP 2

A range of different methods is available in climate impact assessment, some of which will, depending on the circumstances, be more appropriate than others. A combination of methods (such as process-based modelling combined with expert judgement) may be optimal in many cases. Each method, however, requires careful testing and, in the case of models, calibration. This is the subject of the next chapter.

Chapter 6

THE THIRD STEP:

Testing the Method

Following the selection of the assessment methods, it is important that these are thoroughly tested in preparation for the main evaluation tasks. There are many examples of studies where inadequate preparation has resulted in long delays in obtaining results. Moreover, this step provides an opportunity to refine goals and evaluate constraints that may have been overlooked (for example, in selecting off-the-shelf models). Three types of analysis may be useful in evaluating the methods: feasibility studies, data acquisition and compilation, and model testing.

FEASIBILITY STUDIES

One way of testing some or all of the methods is to conduct a feasibility or pilot study. This usually focuses on a subset of the study region or sector to be assessed. Case studies such as these can provide information on the effectiveness of alternative approaches, of models, of data acquisition and monitoring, and of research collaboration.

Feasibility studies are most commonly adopted as a preliminary stage of large multidisciplinary and multisectoral research projects. Here, effective planning and scheduling of research relies on the assurance that different research tasks can be undertaken promptly and efficiently. Several approaches can be suggested for conducting feasibility studies:

- evaluation of available information;
- vulnerability analysis;
- geographical zoning;
- microcosm studies;
- response surfaces;
- analogue studies.

Evaluation of available information

The importance of identifying the main data requirements in an impact assessment has already been stressed in Chapter 4. In addition, a review of the published literature should always be undertaken in order to pro-

Table 6.1 *Hypothetical example of a qualitative screening analysis to assess the vulnerability of human settlements to climatic variations*

Settlement	Climate Vulnerability Rating (examples)			
	Drought effects on agriculture	Drought effects on water supply	Flooding effects on buildings	Rural–urban migration
Villages				
<500 people	1, U	2, U	4, L	2, L
Market towns				
500–1000 people	2, U	2, U	4, L	2, U
City A	4, U	3, U	2, L	1, L

Ratings: 1 = Large or very important; 5 = Trivial; L = Likely; U = Unlikely
Source: Adapted from Scott, 1993.[1]

vide a background understanding of the study region, system or activity being investigated, to examine parallel or related studies that have been completed; to obtain new ideas on methods; to locate new sources of data; and to identify possible research collaborators.

Vulnerability analysis

Assuming that the general sector or sectors of interest have already been identified, a useful first step in defining the specific exposure units to be studied is to conduct a vulnerability analysis. This is a qualitative screening procedure which classifies climatic vulnerability in a matrix format. Different exposure units within the sector(s) are entered on one axis, classified, for instance, by type or by scale. On the other axis some effects of climate are categorized, for example, by type of climatic event, by possible future climate changes, or

by a combination of these. Qualitative ratings are then assigned to each cell in the matrix, indicating both the likely size of the effect and its probability of occurrence. These estimates can be made using whatever information there is available – for instance, from previous studies, expert opinion, literature review or simple quantitative assessments (see below). An example is presented of a vulnerability rating for human settlements in Table 6.1, in which different sizes of settlement are assessed for their vulnerability to different types of impact (from drought, flooding, rural–urban migration, etc).[2]

Assessments of this kind can assist in selecting the most appropriate locations for case studies, or the most suitable exposure units. Figure 6.1, for example, illustrates the worldwide variation in levels of food security under current climate. The generally low levels of security in Africa, together with a quite heterogeneous

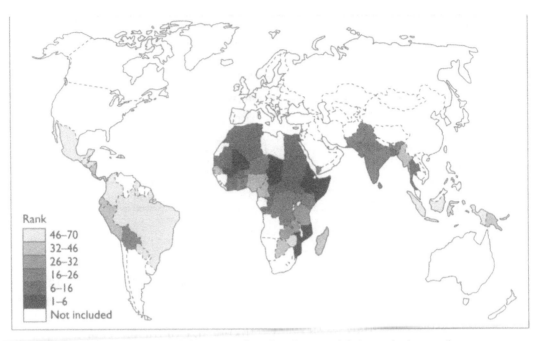

Note: The index is the rank order of 69 countries based on the sum of their standard scores for component indices of national self-reliance and household food access. Countries without shading were not included in the analysis. Low ranks indicate high vulnerability.

Source: Downing, 1992.[3]

Figure 6.1 *Index of food security in developing countries*

pattern, suggest that a study of the potential effect of climate change on global future food security (of the kind reported in Chapter 8) should opt for a worldwide spread of case studies with a relatively dense coverage in Africa.[4]

In a similar way, other forms of vulnerability analysis may help in selecting the most appropriate geographical scale of analysis, the time frame of the study and hence the projection horizon for different scenarios and the types of assessment tools that are appropriate (including models, survey methods, visualization tools and decision support systems).

However, caution should be exercised in interpreting too much from a preliminary assessment of this kind, and this type of procedure should not be regarded as a substitute for an indepth assessment.

Geographical zoning

One method of targeting appropriate areas in which to conduct an impact study is to use simple, large-area geographical zonation. This has been widely used in assessing agricultural and ecosystem impacts, but is potentially applicable in other sectors. It commonly involves the calculation of simple bioclimatic indices, which combine information on climate, soils and topography into measures of suitability for plants and animals. An example of this zoning has been discussed in Chapter 5 in relation to climatic suitability for insect vectors of disease (Figure 5.6). Screening analysis of this kind can indicate zones where the effects of climate change may be most significant. These areas can then be targeted for more detailed analysis (for example, using process-based models that simulate the response of insects to changes in climate, or field experiments to measure their responses).

Microcosm case studies

In studies where there is likely to be a heavy reliance on a specific type of analysis (for example, model-based, experimental or survey-based), or where data requirements are uncertain, it can be instructive to conduct a small-scale pilot study under conditions representative of those anticipated in the main study. These 'microcosm' case studies allow different analysis tools to be selected, tested and evaluated. In addition, they can assist in identifying the personnel required to carry out research. They also offer researchers some experience in addressing problems they are likely to encounter in the main project. For instance, in a project on the ski industry, a representative ski resort might be chosen as a pilot case study; for a study of coping strategies for drought in an agriculturally based, rural, subsistence economy, a representative village might be selected for a pilot survey and analysis.

Response surfaces

A growing number of detailed climate impact studies are being reported for different sectors and from many regions of the world. While these frequently make use of sophisticated analytical methods or models, their results can often be summarized more simply using generalized response surfaces. For example, hydrological models may have been applied to different points in a river catchment and run for different climate change scenarios. The hydrological responses can be complex, but it may still be possible to separate out the most important responses to climate as simple empirical relationships (for example, relating river discharge to monthly precipitation).

Where simple relationships of this kind can be identified from previous studies, there may then be an opportunity to apply them to similar regions in the new study, to provide a preliminary assessment of possible responses to climate change. The use of response surfaces in studying system sensitivities to climate change is discussed further in Box 6.1.

BOX 6.1
SENSITIVITY STUDIES AND RESPONSE SURFACES

One method of representing a range of future climates is to develop response surfaces that depict (usually in two or three dimensions) the response of an exposure unit to all relevant and plausible combinations of climatic forcings. There are numerous derived variables of practical importance, such as soil moisture, runoff, frost frequency, accumulated temperature or flood frequency and return periods, that depend in a non-linear fashion on more fundamental climatological variables such as temperature, precipitation, cloud cover and windspeed (Pittock, 1993).[5]

The figure shows a response surface for snowcover duration (in days), as simulated by an impact model, as a function of changes in temperature and precipitation for a location near Falls Creek in Victoria, Australia (Whetton et al, 1992).[6] The '+' symbol marks the duration for the present climate (no change) and the rectangle represents durations possible for a range of future climates given in regional scenarios produced for 2030.

Clearly, alternative climate change scenarios (for instance, for more distant time horizons, or representing updated knowledge) can be applied readily to a plot of this kind. Moreover, the response surface clearly indicates those combinations of temperature and precipitation change that would be required to produce a given (perhaps critical) response (such as a critical threshold of snow duration below which investment in snow removal equipment for transportation could not be economically justified).

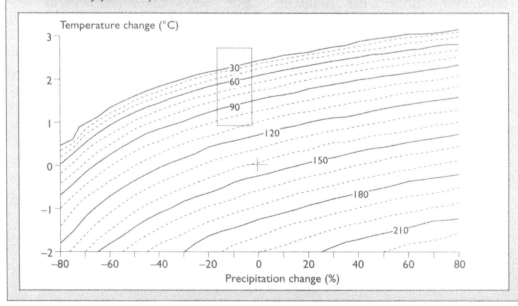

Analogue studies

Another method of rapidly evaluating the likely climate sensitivity of an exposure unit is to identify analogues of possible future conditions. These have already been discussed in the context of a full impact assessment (see Chapter 5) – here they are used as a screening device. These might be regional analogues, where the present-day climate and its effects on an exposure unit are thought to be comparable to possible future conditions in the study region. This is an attractive device for illustrating the possible extent of future climate change, as well as offering useful information on the conditions experienced under the analogue climate. Alternatively, they could be temporal analogues, which identify climatic events and their impacts in the past as analogues of events which could occur again in the future, possibly with an altered frequency under a changed climate.

DATA ACQUISITION AND COMPILATION

An essential element in all climate impact assessment studies is the acquisition and compilation of data. Quantitative data are required both to describe the temporal and spatial patterns of climatic events and their impacts and to develop, calibrate and test predictive models. Four main types of data collection can be identified: empirical compilation, objective survey, targeted measurement and monitoring.

Empirically compiled data

Empirical compilation of evidence (both quantitative and qualitative) from disparate sources is the mainstay of most historical analysis of past climate–society interactions. The data are pieced together to produce a chronology of events, which can then be used to test hypotheses on the effects of past climate,[7] or simply as a qualitative description of past events.[8]

A well-known quantitative reconstruction of this kind is the compilation of the longest instrumentally based record of air temperature. This was assembled by Manley as a table of monthly means for central England from 1659 onwards[9] and is routinely updated on a daily basis.[10] The series was derived from the average of data recorded in Oxford and in Lancashire from 1815. From 1771 to 1815 it was built up by averaging the departures for each month at a number of inland stations whose records are sufficiently long to be 'bridged' into the later run of data. Before 1771 data were bridged into the record from 1726 onwards largely from the midlands of England, with additions from London (from 1723), Upminster (from 1699) and from a variety of scattered observation points from about 1640. In recent years temperatures from Rothamsted (south-east England), Malvern (south midlands), and an average of Ringway (Manchester) and Squires Gate (Blackpool) in north-west England have been used.[11]

The temperature record for central England is shown in Figure 6.2. It is thought to be reasonably accurate from about 1720 onwards, and probably represents an acceptable approximation from 1660. As a result, we have a picture of a general rise in mean temperatures from the late 17th century to the present, with perhaps a slight dip towards the end of the 19th century. Within this general trend lie shorter-term variations of temperature, characterized by runs of perhaps four or five warm or cool years. Such a record provides an invaluable measure of past climatic variability which has enabled scholars to re-examine the role of climate in affecting the social and economic history of England. It also offers a perspective on the extent of projected climate change relative to past variations.

Objective survey

Objective survey utilizes established procedures to collect data from contemporary sources (the information itself may relate to the present or the past). Such survey material may represent either a subset of a population (for instance, a sample of plant species at randomly selected locations within given ecological zones, to be related to climate at the same localities) or the complete population (for example, a regional register of all reported illnesses during a given period that can be related to extreme weather conditions). The tools employed in data acquisition include use of government statistical sources, different methods of questionnaire survey and biological survey techniques. The types of studies reliant on this kind of information include most social impact assessments,[12] studies of perception[13] and studies of biophysical impacts where quantitative data are lacking (for example, data on village-level drought effects on agriculture).[14] Studies of villages during drought in Kenya have revealed, for example, a wide array of coping strategies that are employed to maintain household income (see Table 9.2 in Chapter 9).

Targeted measurement

This refers to the gathering of unique data from experiments where data and knowledge about vital processes or interactions are lacking. This type of measurement is especially important in considering the combined effects of future changes in climate and other environmental factors, combinations which have never before been observed. In many cases these data offer the only opportunity for testing predictive models (for example, observations on the effects of enhanced atmospheric CO_2 on plant growth).

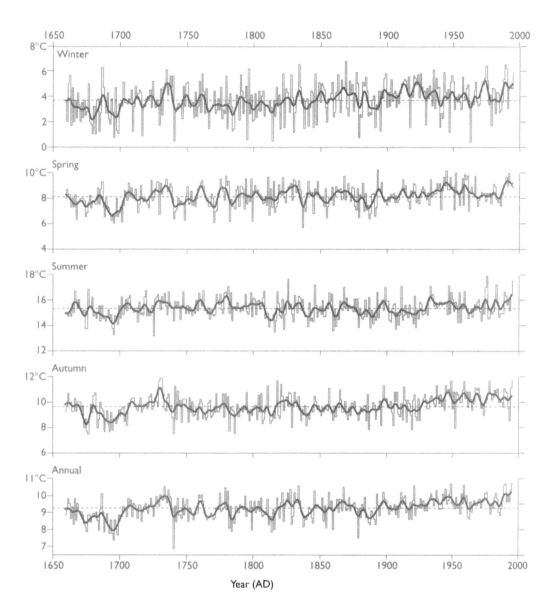

Note: Horizontal dashed lines are the long-term average and bold lines are a filter designed to accentuate decadal variations.

Source: Jones and Hulme, 1997,[15] based on the series constructed by Manley, 1974,[16] and updated by the UK Hadley Centre.[17]

Figure 6.2 *Seasonal and annual temperatures for central England, 1659–1995*

Monitoring

Monitoring is a valuable source of information for climate impact assessment. Consistent and continuous collection of important data at selected locations is the only reliable method of detecting trends in climate itself, or in its effects. In most cases, impact studies make use of long-term data from existing sources (for example, observed climatological data or remotely sensed data). However, in some projects monitoring may form the central theme of research. In these, it is important to consider aspects such as site selection, multiple-uses of single sites, design of measurements and their analysis. It should be noted that there are numerous national and international monitoring programmes. It is important that results from such programmes are made available to impact researchers for assessment studies.

DATA SOURCES

Impact assessments are often hampered by the failure to assemble appropriate data for a given task. This can be due to many causes, including a failure to locate where data are held, bureaucratic delays in the release of data, particularly across national boundaries, and the high cost of obtaining some types of information. Some of these issues are discussed by Hulme[18] with reference to the accessing of climatological data for European countries. The problem of data acquisition is particularly relevant in developing countries.

Where existing data are concerned, government offices often hold valuable data for impact assessment, although the custodians of such data may not be aware of its special relevance. In many cases, data held by the central statistical office of a country are often limited in subject matter and regional coverage, and researchers may need to access data archived in departmental or regional offices. In some cases, national or regional data may be more easily accessible from international organizations. The UNEP GEMS 'Harmonization of Environmental Monitoring' disk is a useful guide to data banks held by various organizations. Some important international sources of data are listed in Appendix 2.

MODEL TESTING

The testing of predictive models is, arguably, the most critical stage of an impact assessment. Most studies rely almost exclusively on the use of models to estimate future impacts. Thus, it is crucial for the credibility of the research that model performance is tested as rigorously as possible. Standard procedures should be used to evaluate models, but these may need to be modified to accommodate climate change. Two main procedures are recommended – validation and sensitivity analysis – and these should always precede more formal impact assessment.

Validation involves the comparison of model predictions with real world observations to test model performance.* The validation procedures adopted depend to some extent on the type of model being tested. For example, the validity of a simple regression model of the relationship between temperature and grass yield should necessarily be tested on data from additional years that are not used in the regression. Here, the success of the model is judged by its outputs, namely the ability to predict grass yield (see Figure 6.3a). Conversely, a process-based model might estimate grass yield based on basic growth processes, which are affected by climate, including temperature. Here, the different internal components of the model (such as plant development and water use) as well as final yield each need to be compared with measurements (see Figure 6.3b).

One problem often encountered in applying process-based models in less developed countries (LDCs) is that the models, while extensively validated in the data-rich developed world, are found to be ill-suited or poorly calibrated for use under the different conditions often experienced in LDCs. A lack or paucity of data for validation may mean that a data-demanding model cannot be used under these circumstances and that a model less dependent on detailed data may be more appropriate.

Climate change introduces some additional problems for validation, since there may be little local data that can be used to test the behaviour of a modelled system in conditions resembling those in the future. Process-based models ought, in theory, to be widely applicable and anyway should be tested in a range of environments. There are fewer grounds, however, for extrapolating the relationships in empirical-statistical models or in most economic models outside the range of conditions for which they were developed.

Sensitivity analysis evaluates the effects on model performance of altering the model's structure, parameter values, or values of its input variables. Extending these principles to climate change requires that the climatic input variables to a model be altered systematically to represent the range of climatic conditions likely to occur in a region. The response surface presented in Box 6.1 is an example of a sensitivity analysis, in this case of snowcover in relation to temperature and precipitation changes.

Through sensivity analysis, information can be gained on:

- the sensitivity of the outputs to changes in the inputs; this can be instructive – for example, in assessing the confidence limits

---

* The term validation is commonly used to describe the testing of model performance. A model can be regarded as a hypothetical representation of reality and, in accordance with the classical scientific method, progress in science occurs by disproving or invalidating hypotheses and formulating new ones. In this sense it is impossible to validate a model.

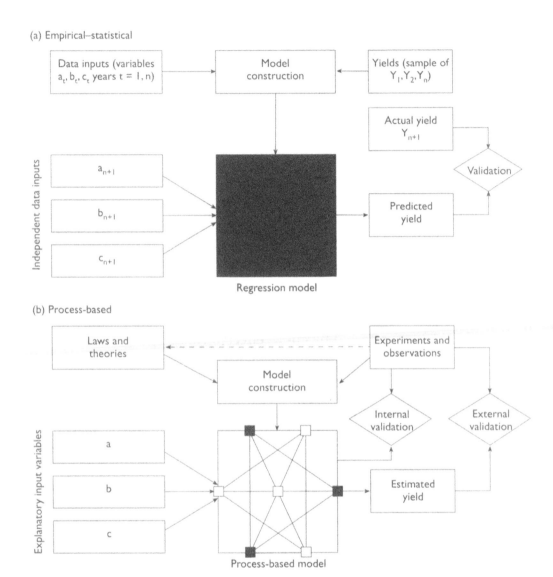

(a) Empirical–statistical

(b) Process-based

Source: Modified from Carter et al, 1988.[19]

Figure 6.3 *The construction, operation and validation of (a) an empirical–statistical (black-box) crop yield model and (b) a process-based crop growth model*

surrounding model estimates arising from uncertainties in the parameter values;
- model robustness (the ability of the model to behave realistically under different input specifications, and the circumstances under which it may behave unrealistically);
- the full range of potential model

application (including its transferability from one climatic region to another, and the range of climatic inputs that can be accommodated).

CONCLUSIONS FOR STEP 3

Not all studies of climate impact will require custom-built methods of analysis. It is sometimes possible to adopt those that have been developed for other purposes, thus saving considerable time and expense. But whether the methods are borrowed or newly developed, they will require careful testing for their feasibility. All too frequently, for example, models have been taken off the shelf from previous studies and misapplied to problems for which they are inappropriate.

Chapter 7

THE FOURTH STEP:

Developing the Scenarios

In this chapter, aspects of the selection and construction of scenarios[*] for use in climate impact assessment are outlined. At the outset, it is important to recognize that the environment, society and economy are not static. Environmental, societal and economic changes will continue, even in the absence of climate change. In order to estimate the environmental and socio-economic effects of climate change, it is necessary to separate them from unrelated, independent environmental and socio-economic changes occurring in the study area. Thus, there is a need first to develop baselines that describe current climatological, environmental and socio-economic conditions (see Box 7.1). Secondly, it is necessary to project future environmental and socio-economic conditions over the study region in the absence of climate change. Projections should take into account, as far as is possible,

autonomous adjustments (see Chapter 9) which are likely to occur in response to changes in these conditions. The third task is to construct scenarios of future climate and scenarios of environmental and socio-economic conditions that are consistent with these. The impacts of climate change are then computed as the difference between future conditions estimated under climate change and future conditions in the absence of climate change. Thus, development of baselines representing current and projected conditions in the absence of climate change is a key and fundamental step in assessment.

ESTABLISHING THE PRESENT SITUATION – THE CURRENT BASELINE

In order to provide reference points for the present-day with which to compare future projections, three types of

---

[*] A scenario is 'a coherent, internally consistent and plausible description of a possible future state of the world'.

BOX 7.1

DIFFERING BASELINES FOR CLIMATE IMPACT ASSESSMENT

The three figures illustrate schematically how differing degrees of realism in assessing impacts result from alternative assumptions about the baseline and from considering various types of adaptation. In Figure A, impacts in the year 2050 (I_1) are portrayed as the cumulative effects of future climate change on an exposure unit, assuming a fixed baseline (no concomitant changes in the environmental, technological, societal and economic conditions relative to the present). This unrealistic, though readily applicable, representation of the future is characteristic of many early climate impact assessments.

Figure B shows how realism is introduced if impacts of future climate change are evaluated relative to a future baseline without climate

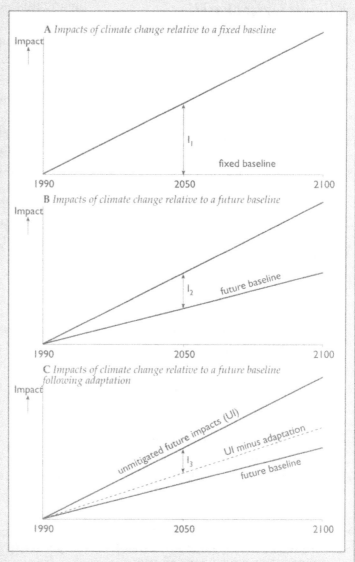

change. The impact relative to the future baseline may be greater or, as is shown in Figure B, less than the impact relative to the fixed baseline (I_2). However, this approach still ignores the many adjustments and adaptations that would occur either in expectation of, or in response to, impacts of climate change. These are shown in Figure C where the objective of adaptation is to reduce the negative impacts of climate change (I_3).

current baseline conditions need to be specified: the climatological, environmental and socio-economic baselines.

Climatological baseline

The climatological baseline is usually selected according to the following criteria:

- representative of the present-day or recent average climate in the study region;
- of a sufficient duration to encompass a range of climatic variations, including a number of significant weather anomalies (eg, severe droughts or cool seasons) – such events are of particular use as inputs to impact models, providing a means to evaluate the impacts of the extreme range of climatic variability experienced at the present-day;
- covering a period for which data on all major climatological variables are abundant, adequately distributed and readily available;
- including data of sufficient quality for use in evaluating impacts;
- consistent or readily comparable with baseline climatologies used in other impact assessments.

A popular climatological baseline is a 30-year 'normal' period as defined by the World Meteorological Organization (WMO). The current standard WMO normal period is 1961–90. As well as providing a standard reference to ensure comparability between impact studies, other advantages of using this as a baseline period include:

- The period ends in 1990, which is the common reference year used for climatic, environmental and socio-economic projections by the IPCC.
- It represents the recent climate, to which many present-day human or natural systems are likely to have become reasonably well adapted (though there are exceptions, such as vegetation zones or groundwater levels, that can have a response lag of many decades, or more, relative to the ambient climate).
- In most countries it is the period for which observed climatological data are most readily available, especially computer-coded data at a daily time resolution.

Nevertheless, in selected cases there may still be valid arguments for choosing an alternative baseline period. These include:

- In some countries, there is better access to climatological data from an earlier period (eg 1951–80 or 1931–60).
- In some, though not all, regions more recent periods (particularly those that include the 1980s) may already contain a significant

global warming signal.

- A 30-year period may not be of a sufficient duration to reflect natural climatic variability on a multidecadal timescale, which could be important in considering long-term impacts.

In recent years great efforts have been expended in the development of high-quality baseline climatological data sets. These can be for individual meteorological stations or averaged over a regular grid. Some of these data sets are available through national agencies while others are supranational or global in scope (see Appendix 2).

Current environmental baseline

The environmental baseline refers to the present state of non-climatic environmental factors that affect the exposure unit. It can be defined in terms of fixed or variable quantities. A fixed baseline is often used to describe the average state of an environmental attribute at a particular point in time. Examples include: mean atmospheric concentration of carbon dioxide in a given year, physiographic features, mean soil pH at a site or mean sea level (which might be expected to alter as a result of future climate change). A fixed baseline is especially useful for specifying the 'control' in field experiments (for instance, the effects of CO_2 on plant growth).

It may be necessary to represent variability in the baseline in order to consider the spatial and temporal fluctuations of environmental factors and their interactions with climate. For example, in studies of the effects of ozone and climate on plant growth, it is important to have information both on the mean and on peak concentrations of ozone under present conditions.

Current socio-economic baseline

The socio-economic baseline describes the present state of all the non-environmental factors that influence the exposure unit. The factors may be geographical (such as land use or communications), technological (pollution control, water regulation), managerial (forest rotation, fertilizer use), legislative (water use quotas, air quality standards), economic (income levels, commodity prices), social (population, diet), or political (land set-aside, land tenure). All of these are liable to change in the future, so it is important that baseline conditions of the most relevant factors are noted, even if they are not required directly in impact studies.

TIME FRAME OF PROJECTIONS

A critical consideration for conducting impact assessments is the time horizon over which estimates are to be made. Three elements influence the time horizon selected: the relevance to timescales of impact and adaptation, the limits of predictability and the compatibility of projections.

Relevance to timescales of impact and adaptation

In Chapter 3 we emphasized that different exposure units are sensitive to different weather events and thus to different timescales of climate change. Furthermore, the time required to respond to a given change in climate may vary greatly: it may take only a few weeks to undertake an actuarial assessment of altered insurance risk and to upgrade premiums to match increased flood hazard, but developing new crop cultivars and encouraging their adoption in farming systems may take 20 years. And to design and build large reservoirs and irrigation systems may take more than 30 years. Finally, the planning horizon of different activities also varies: from two to three years for a new crop rotation, to more than 100 years for a new water reservoir. The extent to which these timescales differ (for climate predictions, for 'impacting' weather events and for the impacts themselves) is illustrated in Figure 7.1.

Limits of predictability

Due to the large uncertainties associated with long-term estimates of future climate and to constraints on computational resources, most GCM simulations have been conducted for periods of up to about 100 years into the future, although a few have also been made over longer time periods of several centuries. For this reason,

the outer horizon commonly adopted in impact studies has been the year 2100. However, within the context of the UN Framework Convention on Climate Change, there is a requirement to specify 'dangerous' levels of anthropogenic interference (interpreted as GHG concentrations). Such levels, and the climate changes associated with them, may not be reached until after 2100, so there may be a need for impact assessments over periods extending beyond this conventional time horizon.

Of course, long timescale projection periods may be wholly unrealistic for considering some impacts (for example, in many economic assessments where projections may not be reliable for more than a few years ahead). On the other hand, if the projection period is too short, then the estimated changes in climate and their impacts may not be easily detectable, making it difficult to evaluate policy responses. Caution must be exercised, therefore, in ensuring that the projection period is both relevant for policy but also valid within the limitations of the approach.

Compatibility of projections

It is important to ensure that future climate, environment and socio-economic projections are mutually consistent over space and time. Many of these are, in any case, intimately related. For instance, changes in greenhouse gas concentrations are related to economic activity and

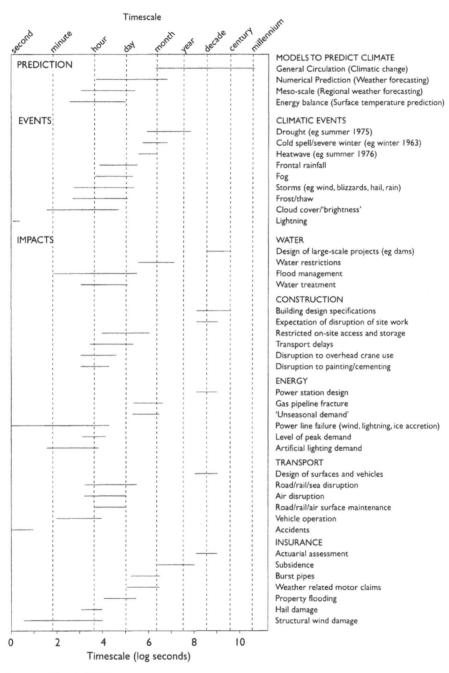

Source: Parry and Read, 1988.[1]

Figure 7.1 *Comparison of the temporal limits of climatic prediction ability with the temporal resolution of climatic events and of their impacts on five sectors of UK industry*

resource use, which are themselves a function of increasing human population. A common area of confusion concerns the relative timing of CO_2 increase and climate change. Thus, it should be noted that an equivalent 2 x CO_2 atmosphere – in which the combined effect of CO_2 and other greenhouse gases on the earth's radiation balance is equivalent to the effect of doubling CO_2 alone – does not coincide in time with an atmosphere in which CO_2 levels themselves have been doubled. Moreover, there is a time lag of several decades in the climate response to the radiative forcing (see Box 7.2).[2,3,4] Hereafter the terms '2 x CO_2' or 'doubled-CO_2' imply a radiative forcing equivalent to 2 x CO_2.

PROJECTING ENVIRONMENTAL AND SOCIO-ECONOMIC TRENDS IN THE ABSENCE OF CLIMATE CHANGE – THE FUTURE BASELINE

Future environmental baseline

The development of a baseline describing conditions without climate change is crucial, for it is this baseline against which all projected impacts are measured. It is highly probable that future changes in other environmental factors will occur, even in the absence of climate change, which may be of importance for an exposure unit. Examples include deforestation, changes in grazing pressure, changes in groundwater level and changes in air, water and soil pollution. Official

projections may exist to describe trends in some of these (such as groundwater level), but for others it may be necessary to use expert judgement. Most factors are related to, and projections should be consistent with, trends in socio-economic factors (see below). Greenhouse gas concentrations may also change, but they are usually linked to climate (which is assumed unchanged here).

Future socio-economic baseline

Global climate change is projected to occur over time periods that are relatively long in socio-economic terms. Over that period it is certain that economy and society will change, even in the absence of climate change. One of the most difficult aspects of establishing trends in socio-economic conditions is forecasting future demands on resources of interest. Simple extrapolation of historical trends without regard for changes in prices, technology or population will often provide an inaccurate base against which to measure impacts.

Official projections exist for some of these changes, since they are required for planning purposes. Projections vary in their time horizon from several years (for example, economic growth and unemployment), through decades (urbanization, industrial development, agricultural production) to a century or longer (population). Reputable sources of such projections include the United Nations, the Organization for

Economic Cooperation and Development (OECD), the World Bank, the International Monetary Fund (IMF) and national governments (see Appendix 2). Some examples of recent global projections based on the IS92 scenarios described in Chapter 2 are summarized in Table 7.1. Nevertheless, many of these are subject to large uncertainties due to political decisions (for instance, international regulations with respect to production and trade) or unexpected changes in political systems (for example, in the USSR, eastern Europe and South Africa during the early 1990s).

Other trends are more difficult to estimate. For example, advances in technology are certain to occur, but their nature, timing and effect are almost impossible to anticipate. In some sectors, it is possible to identify trends in past impacts as attributable to the effects of technology (for example, on health or crop yields). In these cases, changes in technology can be factored in either by examining past trends in resource productivity or by expert judgement, considering specific technologies that are on the horizon and their probable adoption rates, or by a combination of these.

PROJECTING FUTURE CLIMATE

In order to conduct experiments to assess the impacts of climate change, it is first necessary to obtain a quantitative representation of the changes in climate themselves. No method exists yet of providing confident predictions of future climate. Instead, it is customary to specify a number of

Table 7.1 *Scenarios developed and used by the IPCC*

Names of IPCC Scenarios	1990	Scenario for 2100					
		IS92a	IS92b	IS92c	IS92d	IS92e	IS92f
Population (billion)[a]	5.252	11.3	11.3	6.4	6.4	11.3	17.6
Economic growth rate (annual GNP (%))[a]	–	2.3	2.3	1.2	2.0	3.0	2.3
CO_2 concentration (ppmv)[b]	355	733	710	485	568	986	848
Global mean annual temperature change (°C)[b,c]	0	2.47	2.40	1.53	1.91	2.84	2.92
Range (°C)[b,d]	–	1.62–3.75	1.57–3.66	0.97–2.44	1.23–2.99	1.89–4.26	1.93–4.40
Sea level rise (cm)[b,c]	0	45	45	33	38	50	51
Range (cm)[b,e]	–	14–85	13–85	7–68	10–76	17–92	17–95

a Leggett et al (1992).[5] *b* Based on 'best estimate' assumptions given in Wigley and Raper (1992)[6] with CO_2 fertilization feedback included, but using a version of MAGICC (May 1993) giving different values from those reported by Wigley and Raper. *c* Assumes a mid-range climate sensitivity of 2.5°C. *d* Values for low (1.5°C) and high (4.5°C) climate sensitivity. *e* Subjective 10% and 90% confidence levels.

Source: Carter et al, 1994.[7]

BOX 7.2

THE RELATIONSHIP OF EQUILIBRIUM AND TRANSIENT WARMING TO INCREASES IN CO_2 AND IN EQUIVALENT CO_2

The figure below is based on simulations with the MAGICC model (see Table 7.1) of the 'best estimate' of global mean annual temperature change under the IS92a emissions scenario produced for the IPCC,[8] assuming no negative forcing due to sulphate aerosols. It illustrates three important points that are a frequent source of confusion and misunderstanding among impact analysts.

(1) Projected doubling of atmospheric CO_2 occurs at different times depending on the selection of a baseline. Climatologists often refer to pre-industrial CO_2 levels (shown in the figure as a concentration of 279 ppmv in the year

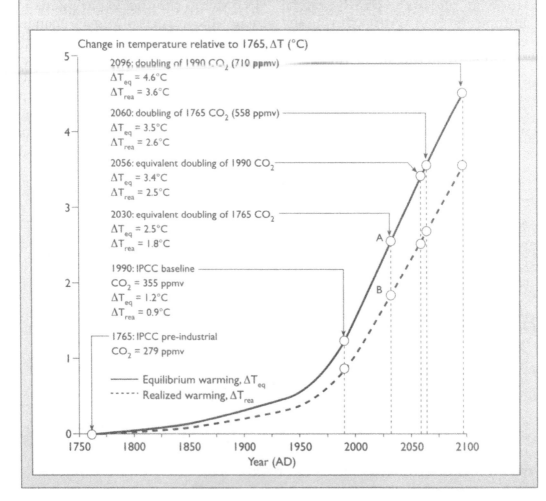

Change in temperature relative to 1765, ΔT (°C)

2096: doubling of 1990 CO_2 (710 ppmv)
ΔT_{eq} = 4.6°C
ΔT_{rea} = 3.6°C

2060: doubling of 1765 CO_2 (558 ppmv)
ΔT_{eq} = 3.5°C
ΔT_{rea} = 2.6°C

2056: equivalent doubling of 1990 CO_2
ΔT_{eq} = 3.4°C
ΔT_{rea} = 2.5°C

2030: equivalent doubling of 1765 CO_2
ΔT_{eq} = 2.5°C
ΔT_{rea} = 1.8°C

1990: IPCC baseline
CO_2 = 355 ppmv
ΔT_{eq} = 1.2°C
ΔT_{rea} = 0.9°C

1765: IPCC pre-industrial
CO_2 = 279 ppmv

—— Equilibrium warming, ΔT_{eq}
------ Realized warming, ΔT_{rea}

Year (AD)

1765) as a baseline to examine effects on climate of subsequent CO_2-forcing. In contrast, impact assessors are more likely to favour selecting a baseline from recent years (eg 1990, concentration 355 ppmv) to provide compatibility with other baseline environmental or socio-economic conditions of importance in impact assessment.

(2) Projected doubling of CO_2 alone occurs significantly later than a doubling of equivalent atmospheric CO_2, where all greenhouse gases are considered. Hence, the doubling date for 1765 CO_2 (2060; 558 ppmv) occurs 30 years later than the equivalent doubling date (2030). Similarly, doubling of 1990 CO_2 to 710 ppmv is projected at 2096, whereas equivalent doubling occurs at 2056.

(3) The actual or 'realized' warming at a given time in response to GHG-forcing (as depicted in transient-response GCM simulations) is less than the full equilibrium response (as estimated by $2 \times CO_2$ GCM simulations), owing to the lag effect of the oceans. These effects can be simulated at a global scale by MAGICC (curves in figure).[9] Thus, at the time of equivalent doubling of 1765 CO_2 (2030), the equilibrium warming relative to 1765 is 2.5°C (point A in figure), while the realized warming is only 1.8°C (point B).

plausible future climates. These are termed 'climatic scenarios' and they are selected to provide information that is:

- straightforward to obtain and/or derive;
- sufficiently detailed for use in regional impact assessment;
- simple to interpret and apply by different researchers;
- representative of the range of uncertainty of predictions;
- spatially compatible, such that changes in one region are physically consistent with those in another region and with global changes;
- mutually consistent, comprising combinations of changes in different variables (which are often correlated with each other) that are physically plausible.

Several types of climatic scenario have been used in previous impact studies. These fall into three main classes: synthetic scenarios, analogue scenarios and scenarios from general circulation models.

Synthetic climatic scenarios

Synthetic scenarios describe techniques where particular climatic elements are changed by a realistic but arbitrary amount (often according to a qualitative interpretation of climate model predictions for a region). Adjustments might include, for example, changes in mean annual temperature of ±1, 2, 3°C, or changes in annual precipitation of ±5, 10, 15 per cent, relative to the baseline climate. Adjustments can be made independently or in combination. Given their arbitrary nature, these are not scenarios in the strict sense, but they do offer useful tools for exploring system sensitivity in impact assessments. In particular, synthetic scenarios can be used to obtain valuable information on:

- The *sensitivity* of the exposure unit to climate change: this can be expressed, for example, as a percentage change in response per unit change in climate relative to the baseline (see, for example, Box 6.1).
- *Thresholds or discontinuities of response* that might occur under a given magnitude or rate of change: these may represent levels of change above which the nature of the response alters (for instance, warming may promote plant growth, but very high temperatures cause heat stress), or responses which have a critical impact on the system (such as windspeeds above which structural damage may occur to buildings).
- *Tolerable climate change*, which refers to the magnitude or rate of climate change that a modelled system can tolerate without major disruptive effects (sometimes termed the 'critical load'); this type of measure is potentially of value for policy, as

it can assist in defining specific goals or targets for limiting future climate change (see Chapter 4).

Analogue climatic scenarios

Analogue scenarios are constructed by identifying recorded climatic regimes, which may serve as analogues for the future climate in a given region. These records can be obtained either from the past (temporal analogues) or from another region at the present (spatial analogues).

Temporal analogues are of two types: palaeoclimatic analogues based on information from the geological record, and instrumentally based analogues selected from the historical observational record, usually within the past century. Both have been used to identify periods when the global (or hemispheric) temperatures have been warmer than they are today. Other features of the climate during these warm periods (such as precipitation, air pressure, windspeed), if available, are then combined with the temperature pattern to define the scenario climate. Palaeoclimatic analogues are based on reconstructions of past climate from fossil evidence, such as plant or animal remains and sedimentary deposits. Three periods have received particular attention: the Mid-Holocene (5000 to 6000 years BP), the Last (Eemian) Interglacial (125,000 BP) and the Pliocene (three to four million years BP).[10] Instrumentally based analogues identify past periods of observed global-scale warmth as an analogue of a GHG-induced warmer world. Maps are constructed of the differences in regional temperature (and other variables) during these periods relative either to long-term averages or to similarly identified cold periods.[11] The main problem with both these types of analogue concerns the physical mechanisms and boundary conditions giving rise to the warmer climate. Aspects of these were almost certainly different in the past compared to those involved in greenhouse gas-induced warming.

Nevertheless, there may be value in identifying weather anomalies from the historical record that can have significant short-term impacts (such as droughts, floods and cold spells). A change in future climate could mean a change in the frequency of such events. For example, several studies have used the dry 1930s period in central North America as an analogue of possible future conditions (see Figure 7.2).[12] Another important anomaly in many regions is the El Niño phenomenon. Changes in the frequency of this event could have significant impacts in many sectors. An extension of this idea is to select 'planning scenarios' representing not the most extreme events, but events having a sufficient impact and frequency to be of concern (for example, a one in ten-year drought event) or consecutive events, whose combined effect may be greater than the sum of individual anomalies.[13]

Spatial analogues identify regions today which have a climate analogous to the study region in the future (for an example, see Figure 5.9). This approach is severely restricted, however, by the frequent lack of correspondence between other non-climatic features of two regions that may be important for a given impact sector (for instance, day length, terrain, soils or economic development).

Climatic scenarios from general circulation models

Three-dimensional numerical models of the global climate system (including atmosphere, oceans, biosphere and cryosphere) are the only credible tools currently available for simulating the physical processes that determine global climate. As indicated in Chapter 2, while simpler models have also been used to simulate the radiative effects of increasing greenhouse gas concentrations, only general circulation models – possibly in conjunction with nested regional models (see below) – have the potential to provide geographically and physically consistent estimates of regional climate change which are required in impact analysis.

General circulation models (GCMs) produce estimates of climatic variables for a regular network of grid points across the globe. Results from about 20 GCMs have been reported to date, and some of the main uncertainties in GCM estimates were described

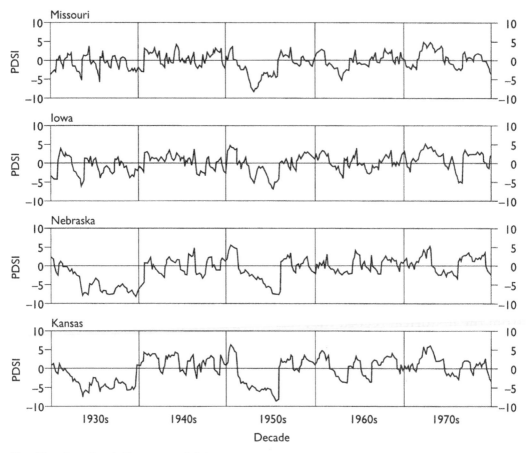

Note: Negative values indicate water deficiency.

Source: Rosenberg et al, 1993.[14]

Figure 7.2 *Palmer Drought Severity Index (PSDI) for the US corn belt, 1938–80*

in Chapter 2. While their ability to simulate the large-scale features of present-day climate is quite good, at a regional scale GCM outputs may sometimes fail to reproduce even the seasonal pattern of current climate. This naturally casts some doubt on the ability of GCMs to produce accurate estimates of future regional climate. Thus GCM outputs should be treated, at best, as broad-scale sets of possible future climatic conditions and should not be regarded as predictions. GCMs have been used to conduct two types of experiment for estimating future climate: equilibrium-response and transient-response experiments.

Equilibrium-response experiments
The majority of experiments have been conducted to evaluate the equilibrium response (new stable state) of

the global climate following an abrupt increase (commonly a doubling) of atmospheric carbon dioxide concentration. Clearly, such a step change in atmospheric composition is unrealistic since increases in GHG concentrations (including CO_2) are occurring continually and are unlikely to stabilize in the foreseeable future. Moreover, since different parts of the climate system respond differently to radiative forcing, they will approach equilibrium at different rates and may never approximate the composite equilibrium conditions modelled in these simulations. Nevertheless, from a modelling point of view these simulations are straightforward to conduct and are useful for model diagnosis and development, as they enable direct intercomparison between models. Results from equilibrium-response experiments have been reviewed by the IPCC in various chapters.[15]

Transient-response experiments
In recent years, more realistic experiments have been conducted with GCMs that simulate the transient-response of climate to a time-dependent change in atmospheric composition.[16] Transient simulations offer several advantages over equilibrium-response experiments. Firstly, the specifications of the atmospheric perturbation are more realistic, involving a continuous time-dependent change in GHG concentrations. Secondly, the representation of the

oceans is more realistic since the simulations couple atmospheric models to dynamical ocean models. Thirdly, transient simulations provide information on the rate as well as the magnitude of climate change, which is of considerable value for impact studies. Fourthly, the most recent transient simulations have also discriminated between the climatic effects of regional sulphate aerosol loading (a negative forcing) and global GHG forcing.

Information from GCMs
The following types of information are available from GCMs for constructing scenarios:

- outputs from a 'control' simulation, which assumes present-day (or, for some recent transient runs, pre-industrial) GHG concentrations, and a 'perturbation' experiment which assumes future concentrations; in the case of equilibrium-response experiments, these are values from multiple-year model simulations for the control and 2 x CO_2 equilibrium conditions; transient-response experiments provide values for the control equilibrium conditions and for each year of the transient perturbation run (eg 1861 to 2100);
- values of surface or near-surface climatic variables for model grid boxes characteristically spaced at intervals of several hundred kilometres around the globe;

- values of air temperature and precipitation (mean daily rate), which are commonly supplied for use in impact studies; data on cloud cover, radiation, windspeed, humidity and other variables are also available from most models;
- data averaged over a monthly time period – in addition, daily or hourly values of certain climatic variables, from which the monthly statistics were derived, may also be stored for a number of years within the full simulation period.

Selecting model outputs
Many GCM simulations have been conducted in recent years and it is not easy to choose suitable examples for use in impact assessments. In general, the more recent simulations are likely to be more reliable as they are based on recent knowledge and they tend to be of a higher spatial resolution than earlier model runs. Several model intercomparison exercises have been conducted in recent years, which provide useful information on model reliability and uncertainties in different regions (Kattenburg et al, 1996, have summarized GCM comparisons for seven world regions: central North America, south-east Asia, the Sahel region in Africa, southern Europe, Australia, northern Europe and east Asia).[17] It is strongly recommended that recent reviews of GCMs are consulted before selection. In addition, several research centres serve as clearing houses for GCM and related information required for developing scenarios (for example, National Center of Atmospheric Research, Boulder, US; the Climatic Research Unit, University of East Anglia, UK; the Commonwealth Scientific and Industrial Research Organization, Australia). Some of these have also developed software for extracting, displaying and comparing information from different GCMs.[18]

Constructing scenarios
Since GCM outputs are not generally of a sufficient resolution or reliability to estimate regional climate even for the present day (via the control run), it is usual for baseline observational data to be used to represent the present-day climate. These are then adjusted by the difference (or ratio) between results for the GCM perturbation experiment and results for the GCM control simulation (for recent transient runs, results representing the present day in the perturbation run are used in place of the control-run results). Differences are usually applied for temperature changes (eg $2 \times CO_2$ minus control) while ratios are commonly used for precipitation change (eg $2 \times CO_2$ divided by control), though differences may be preferred in certain cases.

One of the major problems in applying GCM projections to regional impact assessments is the coarse spatial scale of the gridded estimates, typically spaced at intervals of several hundred kilometres. Several methods have been adopted for developing regional GCM-based scenarios at the subgrid scale.

- The baseline climate in a region or at a site is adjusted based on the GCM-based scenario change for the nearest centre of a grid box.[19] This has the drawback that sites in close proximity but falling in different grid boxes, while having a very similar baseline climate, may be assigned a quite different scenario climate.
- Scenario changes are objectively interpolated to the site or region of interest from nearby grid boxes.[20] This overcomes the problem above, but it introduces a false geographical precision to the estimates.
- Experiments are conducted at higher resolution over the region of interest, either by running a GCM at higher resolution for a limited number of years ('time slice' experiments), or by running a GCM at varying resolution across the globe with the highest resolution over the study region ('stretched grid' experiments), or by running a separate high-resolution model over a limited area, using conventional GCM outputs to provide the boundary conditions for the high-resolution model (the 'nesting' approach).[21] This method of obtaining subgrid-scale estimates (sometimes down

to 50-kilometre resolution) is able to account for important local forcing factors, such as surface type and elevation, which GCMs are unable to resolve. It has the advantage of being physically based, but it is also highly demanding of computer time. For this reason, of the runs that have been conducted to date, very few have been made for a sufficient period of simulated years to allow meaningful climate change statistics to be extracted. Furthermore, the commonest approach, nesting, is still heavily dependent on (uncertain) GCM outputs for its boundary conditions.

- Statistical relationships are established between observed climate at the local scale and at the scale of GCM grid boxes. These relationships are used to estimate local adjustments to the baseline climate from the GCM grid box values.[22] A potential weakness of this method is that it assumes that the relationships between subgrid-scale and large-scale climate will not change under future radiative forcing.

Outputs from GCMs are usually applied as monthly or seasonal adjustments to the baseline climatology in impact assessments, assuming no change in climatic variability between the baseline and future climate. Since changes in climatic variability can have a great effect on

BOX 7.3
DEVELOPING SCENARIOS FROM GCM OUTPUTS

To illustrate how climate change scenarios can be constructed from GCM outputs, let us consider scenarios of summer temperature and precipitation over central India.[23] In the examples presented below, data from 13 GCMs have been interpolated from their original grid box resolutions to a uniform 5° latitude/longitude grid (the region is represented by between 2 and 12 GCM grid boxes). Four 5° grid boxes fall over central India, and values from these have been averaged. In addition, data from the UK Hadley Centre model[24] have been averaged across its 12 model grid boxes over central India.

Defining the baseline climate:

Observations of mean monthly temperature for 1889–1995[25] and monthly precipitation for 1900–95[26] were available as 5° latitude/longitude grid averages for central India. The summer (JJA) precipitation record is shown in Figure A. Long period records like this provide information about inter-annual climatic variability and possible long-term trends in the region. The significance of any future mean change in climate has to be judged in the context of this natural variability.

In this example, the period 1961–90 is adopted as the climatological baseline. The climate of this period is assumed to be representative of the recent regional climatic conditions to which ecosystems and society are adapted. The baseline summer precipitation over central India is depicted as solid bars in Figure B.

A precipitation scenario based on transient GCM outputs

Several recent simulations with coupled ocean–atmosphere GCMs have considered the transient response of climate to a time-dependent change in radiative forcing from pre-industrial times through to the end of the 21st century. Some simulations have considered greenhouse gas (GHG) forcing alone; others have considered the influence of sulphate-aerosol forcing alone or in combination with GHG-forcing. In addition, long-period control simulations, which assume constant pre-industrial levels of greenhouse gases, have also been run to represent the natural variability of climate in the absence of radiative forcing.

Returning to our example, a precipitation scenario for the year 2050 has been constructed based on the Hadley Centre GHG plus aerosols perturbation run. The ratio of mean precipitation during the 30-year period centred on 2050 (2036–65) to mean precipitation simulated for 1961–90 in the perturbation run is 0.89, representing a precipitation decline of 11 per cent (Figure C). Applying this change as an adjustment to each year of the observed 1961–90 baseline yields a scenario time series of precipitation representing the period around 2050 (open bars in Figure B).

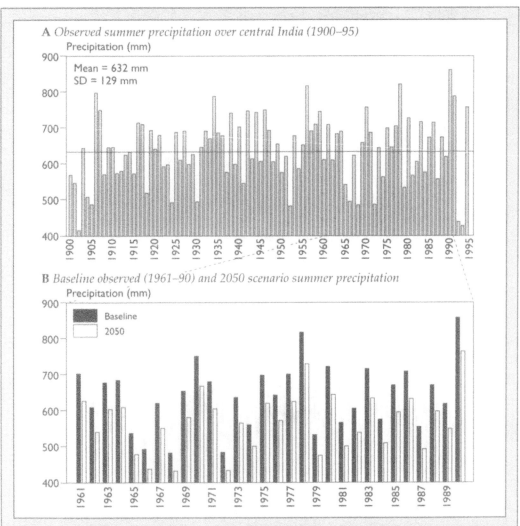

A *Observed summer precipitation over central India (1900–95)*

B *Baseline observed (1961–90) and 2050 scenario summer precipitation*

To assess if this precipitation change describes a true climate change signal, or whether it is merely an expression of the normal, decadal-scale climatic variability, it can be compared to the control simulation (Figure D). In fact, the change (−11 per cent) lies well outside the inter-decadal natural variability for this region described both in the GCM control run and in the observations during 1900–95. The variability of 30-year averaged values through both of these series, expressed by the standard deviation, is about ±2 per cent relative to the mean precipitation (see Figure E, below).

Expressing scenario uncertainties

The scenario presented here is based on just one example of the many GCM experiments conducted to date. It also relies on only one set of assumptions about

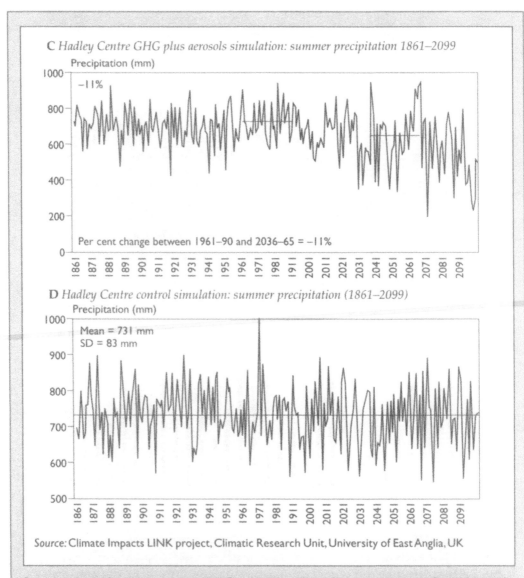

C *Hadley Centre GHG plus aerosols simulation: summer precipitation 1861–2099*

Precipitation (mm)

−11%

Per cent change between 1961–90 and 2036–65 = −11%

D *Hadley Centre control simulation: summer precipitation (1861–2099)*

Precipitation (mm)

Mean = 731 mm
SD = 83 mm

Source: Climate Impacts LINK project, Climatic Research Unit, University of East Anglia, UK

future atmospheric composition. On its own, therefore, it cannot reflect the wide range of uncertainties that exist in regional climate projections. Three main sources of uncertainty can be identified:

- in future GHG and aerosol emissions characterised, for example, by the IS92 scenarios (see Chapter 2) and implying different future levels of atmospheric composition and hence of radiative forcing;
- in global climate sensitivity, with some models simulating greater mean global warming per unit of radiative forcing than others; and
- in regional climate changes, with different GCMs giving different regional estimates of change for the same mean global warming.

Unfortunately, GCMs have not been run for a wide range of emissions scenarios. Instead, analysis of the effect of emissions uncertainties on climate change has commonly been based on simple models, providing estimates of globally averaged temperature change (see Chapter 2). There is evidence from a limited number of GCM runs that the regional pattern of temperature change assuming low GHG emissions is similar to that for higher emissions, although no conclusion can be drawn for precipitation because of its high natural variability.[27] This result provides tentative support for a procedure known as 'scaling', which combines outputs from GCMs and simple climate models. The mean global temperature change under different emissions and climate sensitivity assumptions is first obtained by a simple climate model (see Chapter 2). This can then be used to infer uncertainties in regional climate by scaling the pattern of change from a GCM according to the ratio between the global mean annual temperature change from the GCM and that given by the simple model.

This scaling procedure has been used to compare the outputs from 14 GCM runs for GHG-only forcing (Figure E). Some are equilibrium simulations, others transient simulations. All have been scaled to represent the regional pattern of change over central India for a global mean annual temperature change of 1.4°C, which is assumed to occur by 2050 according to a simple climate model. One of these is the Hadley Centre GHG-only simulation, which represents the 30-year period around 2031, when modelled global mean annual temperature is 1.4°C above the 1961–90 mean. In addition, the changes under the Hadley Centre GHG plus aerosols simulation for 2031 are also shown (Figure E). Interestingly, the warming and slight drying seen with the Hadley Centre GHG-only simulation are actually enhanced in this region with the inclusion of aerosols. This contrasts with the finding globally that average temperature increases are suppressed by aerosol forcing.

The points in Figure E represent changes scaled for the IS92a emissions scenario (see Table 7.1) and a mid-range climate sensitivity (2.5°). The lines passing through the GHG-only points bound estimates of temperature and precipitation change for an extreme range of emissions and climate sensitivity assumptions.[28] In addition, error bars have been attached to the Hadley Centre GHG-only estimate and to the origin that show the natural variability (±2 standard deviations of 30-year smoothed data) represented in the Hadley Centre control run and in the observed data, respectively.

Scatter plots such as that shown in Figure E can be of considerable value to the impact analyst, offering insights into the likely range of future climate changes which can guide sensitivity analysis (see Chapter 6 and Box 6.1). They also provide information for the policy-maker on the uncertainties in future climate projections at the regional level.

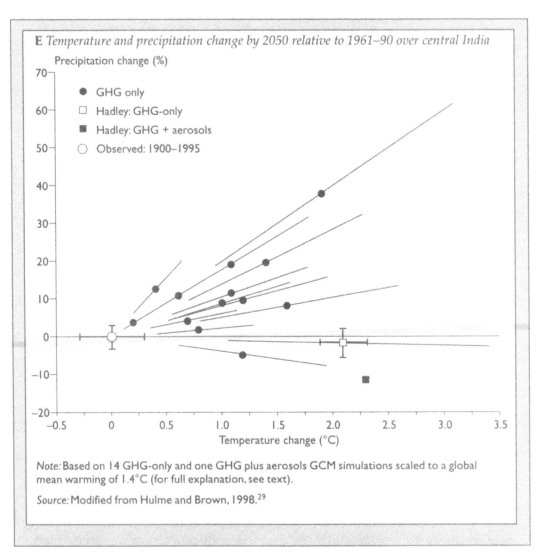

E *Temperature and precipitation change by 2050 relative to 1961–90 over central India*

Precipitation change (%)

- ● GHG only
- ☐ Hadley: GHG-only
- ▨ Hadley: GHG + aerosols
- ○ Observed: 1900–1995

Temperature change (°C)

Note: Based on 14 GHG-only and one GHG plus aerosols GCM simulations scaled to a global mean warming of 1.4°C (for full explanation, see text).

Source: Modified from Hulme and Brown, 1998.[29]

the frequency of extreme climatic events, and their consequent impacts, some recent impact studies have also focused on measures of variability change derived from GCM outputs.[30] Some of these studies make use of stochastic weather generators, which reproduce the main statistical features of the observed weather and can be readily employed to simulate daily time series for a scenario climate with a changed mean or variability.[31] Box 7.3 illustrates in more detail some of the options for producing scenarios using outputs from both equilibrium-response and transient-response experiments.

PROJECTING ENVIRONMENTAL AND SOCIO-ECONOMIC TRENDS WITH CLIMATE CHANGE

Environmental trends with climate change

Future changes in environmental conditions not due to climate factors should already have been incorporated in the development of future environmental baselines (see above). The only changes in these trends to be incorporated here are those associated directly with climate change. The two factors most commonly required in assessments are atmospheric composition and sea-level rise.

Projections of atmospheric composition are important for assessing effects, firstly, on radiative forcing of the climate (both greenhouse gas and aerosol concentrations), secondly, on depletion of stratospheric ozone (CFCs), and thirdly, on plant response (CO_2, tropospheric ozone and compounds of sulphur). In applying them, however, they should be consistent with the projected climate changes. Scenarios for CO_2 concentrations are given in Table 7.1.

Sea-level rise is one of the major impacts projected under global warming. Global factors such as the expansion of sea water and melting of ice sheets and glaciers all contribute to this effect. However, local conditions such as coastal land subsidence, isostatic uplift as well as possible changes in storminess, waves and tides should also be taken into account

in considering the extent of sea-level changes and their regional impacts. In most assessments, the vulnerability of a study region to the effects of sea-level rise will be apparent (for instance, in low-lying coastal zones). Less obvious are some inland locations which may also be affected (for example, through sea water incursion into groundwater). The magnitude of future sea-level rise is still under discussion, but the global estimates given in Table 7.1 (which are consistent with the other changes shown in the table) may serve as a useful basis for constructing scenarios.

Other environmental factors that are directly affected by climate include river flow, runoff, soil characteristics, erosion and water quality. Projections of these often require full impact assessments of their own, or could be included as interactive components within an integrated assessment framework.

Socio-economic trends with climate change

Socio-economic factors that influence the exposure unit may themselves be sensitive to climate change, so the effects of climate should be included. In some cases this may not be feasible (for instance, it is not known how climate change might affect population growth), and trends estimated in the absence of climate change would probably suffice. In other cases, projections can be adjusted to accommodate possible effects of climate (for example, there are quantifiable effects on human health of the interaction between local climate and atmospheric pollution and toxic waste disposal in many urban areas, the causes of which are closely associated with emissions and byproducts of fossil fuel combustion).

There are also many human responses to climate change that are predictable enough to be factored-in to future projections. These are often accounted for in model simulations as feedbacks or 'autonomous adjustments' to climate change and are considered in Chapter 9.

A final factor to consider in projecting socio-economic trends under a changing climate is the effect that various policies designed to mitigate climate change might themselves have on the future state of the economy and society. For example, policies to reduce fossil fuel consumption through higher energy prices might alter the pattern of economic activity, thus modifying the possible impacts of any remaining (unmitigated) changes in climate that occur.

CONCLUSIONS FOR STEP 4

In order to estimate the impacts of future climate change, we need projections not only of the future climate but also of the other social, economic and environmental factors that influence the exposure unit. All of these are uncertain, and in the absence of confident predictions it is necessary to adopt scenarios which reflect plausible future conditions across the range of uncertainty. Impacts of climate change are calculated relative to a reference scenario, which is representative either of the present-day situation (current baseline) or a future situation in the absence of climate change (future baseline).

Chapter 8

THE FIFTH STEP:

Assessing the Impacts

In this chapter we shall consider the different ways in which impact assessments of climate change can best be presented. In general, such estimates are calculated as the differences between those conditions that are projected with climate change and those conditions projected to exist under the future baseline (without climate change; see Box 7.1). A wide array of adaptations is likely to occur in anticipation of such impacts and these need to be incorporated in the impact assessment in order for sufficient reality to be maintained. In practice, however, it is sometimes easier, first, to conduct impact assessments assuming a limited range of adaptation in order to reduce the complexity of the problem. Subsequently, different forms of adaptation can be introduced to the analysis in order to assess their effectiveness in modifying the magnitude and rate of impact. We follow this approach here, describing how estimates of impacts can be presented most effectively assuming limited adaptation; in the next chap-ter we describe the ways in which different forms of adaptation can be evaluated and incorporated in more complex impact assessments. We make a distinction below between quantitative analysis and qualitative description as methods of evaluating and presenting the results of an impact assessment. In reality, of course, these are complementary roads of investigation.

QUANTITATIVE ANALYSIS

Effects at site scale

Most model-based impact assessments provide quantitative measures of impact that express changes from the reference case. Because impact models often require considerable amounts of input data for testing and calibration, their use may be restricted to only a limited number of point locations or sites. Figure 8.1 illustrates tabulated changes in yields of maize in Zimbabwe for three sites under current climate (the base case), under three scenarios of altered weather due to climate change, and under the same scenarios of altered weather together with the direct effects of higher levels of ambient atmospheric CO_2 on plant photosynthesis and water use.[1] The sites have been selected to represent a range of different agro-climatic environments in Zimbabwe and assume, unrealistically, no adaptation. At the bottom of Figure 8.1 is a representation of the effects of two forms of adaptation (fertilizing and irrigation) in reducing the negative effects of the most extreme of the three climate change scenarios.

Some point estimates of impact can be presented most appropriately as changes in frequency of occurrence. To illustrate, one of the ways in which climate change may affect human health is through altered human death rates due to heat stress. The following example is taken from a study of death rates in Shanghai (China) due to high maximum temperatures during summer.[2] Figure 8.2 illustrates three distinct steps in the analysis: firstly, a description of the types of weather condition which are the cause of higher mortality rates in Shanghai; secondly, a comparison of the number of heat-related deaths in an average summer as opposed to the hottest summer currently experienced under present-day climate; and thirdly, estimated future mortality rates under either an arbitrary increase of temperatures (here shown as 2°C and 4°C) or three 2 x CO_2 general circulation model-based climates. This example

nicely encapsulates standard practice in impact assessment by including a definition of climatic conditions which cause the impacts, an assessment of the sensitivity of the exposure unit to current levels of impact, and an estimation of impacts under projected future conditions of altered climate.

Changes in geographical distribution

A common feature of all climate impact assessments is that they have a geographical dimension. Since climate varies over space, its effect on the natural and human environment also varies spatially and its spatial pattern is likely to change as the climate changes. By mapping these altered distributions, it is possible to provide place-specific information for policymakers concerning altered levels of resource availability due to climate change. Our ability to conduct such analyses has been improved with the development of computer-based geographical information systems (GIS), which can be used to store, analyse and depict spatial information.

Analyses of altered distributions have most commonly been used in the context of effects on natural terrestrial ecosystems. A regional example is given in Figure 8.3, which illustrates the likely altered distributions of two plant species in Norway under a 2 x CO_2 climate scenario.[3] The rare alpine species, currently limited to high-altitude and inland locations, may be threatened with extinction (Figure

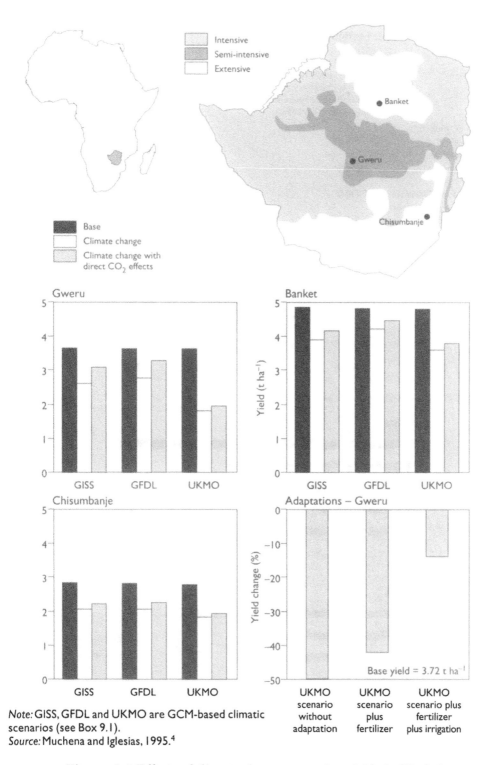

Note: GISS, GFDL and UKMO are GCM-based climatic scenarios (see Box 9.1).
Source: Muchena and Iglesias, 1995.[4]

Figure 8.1 *Effects of climate change on maize yields in Zimbabwe*

Offensive air masses in summer for Shanghai

Air mass type	102	103
Weather	Hottest in the season, light winds from SE, little cloudiness, comparatively high atmospheric pressure (Subtropical Anti-cyclone dominant)	Almost as hot as 102, very light wind from SE at night and from SW by daytime, mostly cloudy, low atmospheric pressure (Maritime Tropical Air Mass)
Mean daily mortality	113.9	116.9
Per cent above overall mean	8.8%	11.2%
Per cent of all summer days	14.2%	17.9%
Days in top 50 highest mortality days (per cent)	17 (34.0%)	30 (60.0%)
Per cent frequency of air mass under GISS 2 x CO_2 scenario	18.4%	32.7%

Heat-related deaths for entire summer

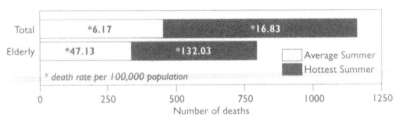

Death rate per 100,000 population

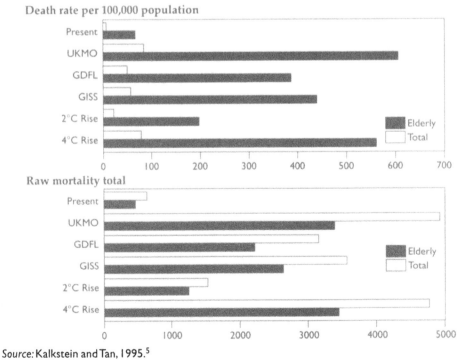

Raw mortality total

Source: Kalkstein and Tan, 1995.[5]

Figure 8.2 *Estimates of future mortality in Shanghai in summer*

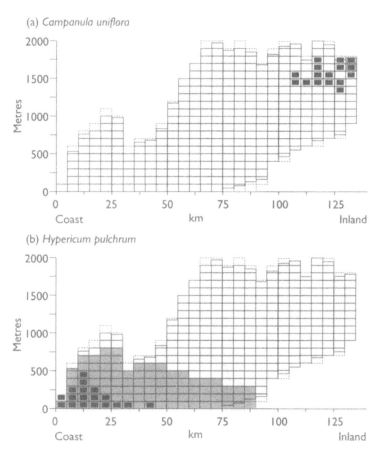

Note: This is a narrow east–west transect, with altitude on the vertical axis and distance from the Atlantic coast on the horizontal axis. Solid squares = current and shaded = projected distribution.

Source: Holten and Carey, 1992.[6]

Figure 8.3 *Altered distribution of two plant species in central Norway under a 2 x CO₂ climate change scenario: (a) for an alpine species* Campanula uniflora *and (b) for a frost-sensitive coastal species* Hypericum pulchrum

8.3a), while a frost sensitive species currently limited to coastal areas in Norway extends its distribution uphill and inland under warmer conditions (Figure 8.3b).

Similar, more broad-scale, geographical approaches have been used to define the likely shift of large-scale vegetation zones or biomes. As an alternative to the complex modelling of plant community behaviour under altered climatic conditions, some studies have sought to define ecoclimatic zones or life zones which bound the climatic and soil conditions associated with a given vegetation or ecosystem type. Figure 8.4 illustrates the potential distribution of major world biomes

Source: Kirschbaum et al, 1996.[7]

Figure 8.4 *Potential distribution of natural vegetation according to the BIOME model:* [8] (a) *for present climate*

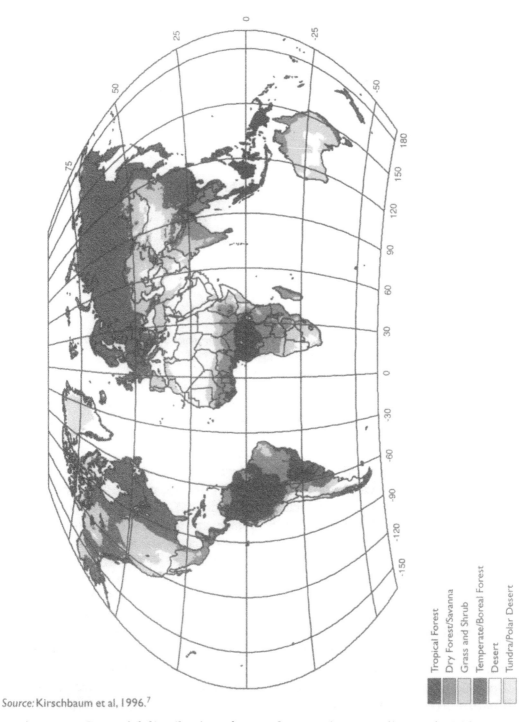

Source: Kirschbaum et al, 1996.[7]

Figure 8.4 *Potential distribution of natural vegetation according to the BIOME model:* (b) *climate as projected by the GFDL GCM*[9]

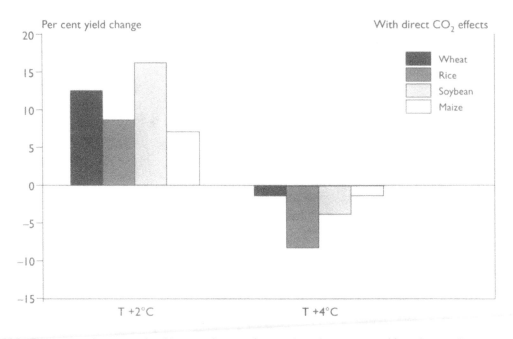

Note: Country results are weighted by contribution of national production to world production. Direct effects of CO$_2$ on crop growth and water use are taken into account.

Source: Rosenzweig et al in Strzepek and Smith, 1996.[10]

Figure 8.5 *Aggregated IBSNAT/ICASA crop model yield changes for +2°C and +4°C temperature increases*

under current climate and under a 2 x CO$_2$ climate change scenarios.[11]

Where several compatible regional studies have been completed, it is possible to expand the scale of estimates from a regional to a global level. For example, the estimates of impact on maize yields in Zimbabwe were one of a set of estimates derived from the IBSNAT/ICASA crop models as part of an international study.[12,13] A description of the IBSNAT/ICASA models is given in Box 5.1. An averaged effect for all countries for the four most important world food crops (wheat, rice, soy-

bean and maize) is shown in Figure 8.5. With the direct effects of CO$_2$ and precipitation held at current levels, average crop yields weighted by national production show a positive response to +2°C warming and a negative response to +4°C warming. Between 2°C and 4°C lies a threshold at which the positive effects of warming (which generally increase photosynthetic rate) are approximately balanced by the negative effects of temperature increases (which generally lead to a shortened period of grain formation and thus lower yields). These averaged results, however, mask important differences between countries. For example, the effects of latitude are such that in Canada a +2°C temperature increase with no precipitation change would result in wheat yield increases, whereas the same changes in Pakistan would result in average wheat yield decreases. Indeed, Figure 8.6 indicates that under all three scenarios of climate change considered in this study, the effects on yield levels of food grains are generally negative at lower latitudes (where higher temperatures and changes in precipitation exacerbate existing problems of water limitation on crop plants), and are generally positive at middle and high-middle latitudes (where high temperatures are either beneficial for crops through a lengthened growing season or, if they are not beneficial, their negative effect on yield is more than compensated for by the direct effects of CO_2). It should be noted, however, that no forms of adaptation (such as changes in crop cultivar, irrigation or altered cropping patterns) have been incorporated in the foregoing analysis. We shall introduce these and other factors in the next chapter.

Changes in risk

Since many impacts from climate change occur through the incidence of extreme weather (such as droughts, floods and hurricanes), an effective way of characterizing a change in climate is through a change in the level of risk. An important point emerging

from such analyses is that probabilities of exceeding given values frequently alter non-linearly, indeed quasi-exponentially, with changes in the mean value due to climate change.[14] For example, the probability of observing a summer in England with temperature conditions similar to that of 1995 (which was the third warmest summer on the 336-year record of temperatures for central England – see Figure 6.2) increases from 13 per thousand under current climate to 333 per thousand under a best-estimate climate change of +1.6°C by the 2050s (Figure 8.7).[15] This assumes that the year-to-year variability of climate remains unaltered in the future. A concurrent increase in variability could conceivably increase this probability. Such a change would have severe implications for agriculture and water resources. Under the same scenario, a very mild winter of the sort observed in 1988–89 would recur on average about every fifth year (twice during the 2050s) compared to three or four times a century at present. By the 2050s decade, the chance of recurrence of a very cold winter in England, such as during 1962–63, is extremely small.

An alternative to portraying altered risk values at a point is to map the spatial shift of given risk values due to climate change. For example, where there is a linear decrease in thermal resources either uphill, due to the lapse rate of temperature, or northwards in mid-latitude areas, due to shorter and less intense growing sea-

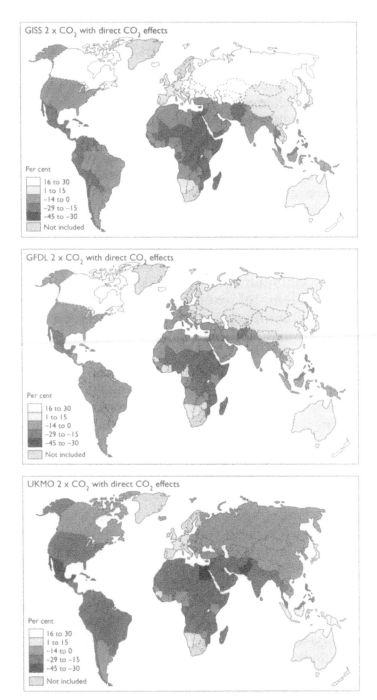

Note: Results shown are averages for countries or groups of countries.

Source: Rosenzweig and Parry, 1994.[16]

Figure 8.6 *Estimated change in average grain yield for GISS, GFDL and UKMO climate change scenarios with direct CO$_2$ effects*

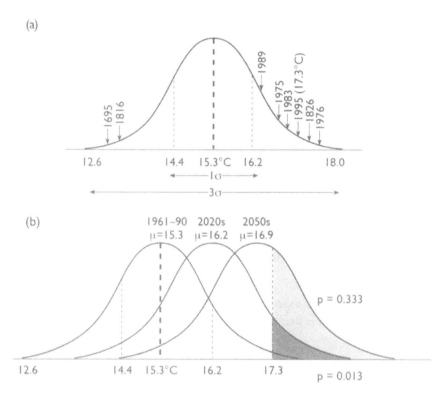

Note: Shaded areas denote changes in approximate exceedance probability for the hot summer of 1995.

Source: UK Department of the Environment, 1996.[17]

Figure 8.7 *Distribution of summer mean temperatures for central England:*
(a) based on the 1962–90 period average and (b) assuming the 1996 Climate Change
Impacts Review Group scenarios for the 2020s and 2050s decades

sons, the risk of there being insufficient warmth to permit crops to reach maturity increases quasi-exponentially.[18] Moreover, the risk of consecutive crop failures can increase even more sharply still (see Figure 8.8). Lines of equal value of probability of crop failure can be mapped and the shift of location of these due to climate change can be plotted (see Figure 8.9).

In those disciplines that inherently deal with probabilites, frequencies and statistical information to characterize natural hazards and to design criteria in order to control structures, the application of risk and uncertainty analyses is a central feature of decision-making and is especially important in hydrology, water management and shore protection.[19]

Costs and benefits

One of the most valuable forms in which impact assessment results can be provided is as costs or benefits.

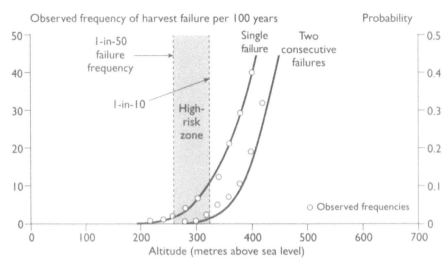

Source: Parry and Carter, 1985.[20]

Figure 8.8 *Actual and assumed frequencies of harvest failure (annual growing degree-days below 970) in southern Scotland, 1659–1981*

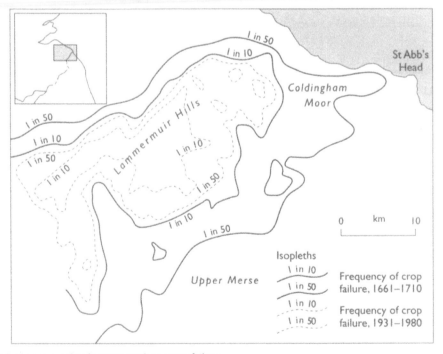

Note: Risk is expressed as frequency of oat crop failure.

Source: Parry and Carter, 1985.[21]

Figure 8.9 *Locational shift of a high-risk zone between the cool (1661–1710) and warm (1931–80) periods in the Lammermuir Hills, south-east Scotland*

Methods of evaluating these range from formal economic techniques, such as cost-benefit analysis, to descriptive or qualitative assessments. Cost-benefit analysis is often employed to assess the most efficient allocation of resources (see Box 8.1). This is achieved through the balancing or optimization of various costs and benefits anticipated in undertaking a new project or implementing a new policy, accounting for the reallocation of resources likely to be brought about by external influences such as climate change. The approach makes explicit the expectation that a change in resource allocation is likely to yield benefits as well as costs, a useful counterpoint to many climate impact studies, where negative impacts have tended to receive the greatest attention.

Whatever measures are employed to assess costs and benefits, they should use common units of value. Thus, for example, where monetary value is ascribed, this is usually calculated in terms of net present value: the discounted sum of future costs and benefits. The choice of discount rate used to calculate present value will vary from nation to nation depending on factors such as the level of economic development, debt stock and social provision. Moreover, the depreciation of capital assets with time, which varies from country to country, should be explicitly considered in the calculations.

In the context of global warming, the relevant costs are those of mitigation and adaptation. These will tend to be expressed in terms of marketed resources such as labour and capital costs. The benefits of mitigation and adaptation are expressed in terms of avoided warming damages. In turn, these damages may show up in terms of market values (lost crops, forest damage etc), and in non-market values (for example, changes in human health, changes in amenity and changes in biodiversity). As far as assessing damages, the distinction between market and non-market values is immaterial: both contribute to human well-being, which is the ultimate yardstick of cost-benefit assessments. In practice, both types of values raise complex issues. Market values may not, for example, represent the true value of resources to a given economy – for example, in the presence of taxes or subsidies or if environmental costs are neglected. In this case, they have to be adjusted to secure their 'shadow' values which measure the cost of the damage to society as a whole. Non-market values have to be elicited by direct and indirect methods such as contingent valuation, hedonic property and wage models or travel cost measures. The resulting estimates should, like shadow market values, reflect the underlying willingness of individuals to pay for the commodity, asset or service that is at risk. There is extensive economic literature on both the methodologies for valuing non-market damages and on empirical estimates (for an overview, see Pearce, 1993).[22]

Box 8.1
COST-BENEFIT ANALYSIS

Cost-benefit analysis has the specific objective of evaluating an anticipated decision or range of decision responses. For example, in considering the costs and
benefits of an adaptation strategy, a cost-benefit analysis might evaluate a question
facing a decision-maker:'Do the benefits of a given level of adaptation outweigh the
costs of its implementation?' The benefits of this action are the avoided damages
(costs) of climate change impacts.

The problem can be illustrated in a simple diagram (see figure). In the lefthand
graph, the degree of adaptation is expressed as two lines: one representing the costs
of adaptation and the other the benefits (avoided costs) accruing from this action.
Both lines show increases with increasing levels of adaptation, but the growth in
costs accelerates, while the growth in benefits diminishes. Characteristically, the costs
of minimal adaptation are small while the benefits are high (eg, at point A), but as the
level of adaptation increases, so the additional or marginal costs increase, while the
marginal benefits decline. These are the slopes of the two lines in the lefthand graph,
plotted as straight lines in the righthand graph. Economic analysis generally con-
cludes that the optimal result is where the marginal cost and marginal benefit of the
change are equal (point B on the graph). To the left of point B further action is beneficial, because the additional (marginal) benefits secured exceed the additional
(marginal) costs. Further adaptation beyond point B produces an unfavourable
cost-benefit ratio (eg, at point C) and is therefore not justified.

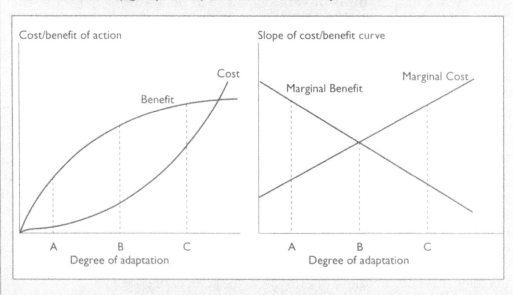

QUALITATIVE ANALYSIS

Some assessments need to be conducted rapidly and at little expense, perhaps to give a preliminary indication of impacts for more detailed subsequent study. In such cases there is frequently resort to expert judgement (see Chapter 4). The success of such evaluations usually rests on the experience and interpretative skills of the analysts, particularly concerning projections of possible future climatic impacts. The disadvantages of subjectivity have to be weighed against the ability to consider all factors thought to be of importance (something that is not always possible using more objective methods such as modelling). The most successful qualitative analyses are those which integrate what is known (as a result of formal research) with what is unknown, and such a mixture of information can assist in this integration. An example, cross-impacts analysis, is illustrated in Box 8.2.

DEALING WITH UNCERTAINTY

Uncertainties can be found at all levels of a climate impact assessment, including the estimation of future emissions of greenhouse gases and aerosols, atmospheric concentrations, changes in climate, future socio-economic conditions, potential impacts of climate change and the effects of adaptation. Uncertainty arises from two main sources: errors and unknowns.

Errors can occur due to inaccurate measurement, lack of data and inadequate parameterization or assumptions. For example, measurement errors can often be quantified by replicating experimental observations or by conducting simultaneous measurements with different instruments or observers.

Unknowns may refer to factors that are not included in an analysis because they are not recognized or understood. Also included in this category are assumptions or estimates to which measures of confidence cannot be attached. Examples include the scenarios described in Chapter 7.

The maximum range of uncertainty is the product of the individual uncertainties, and where quantification of uncertainties is possible, it is common to employ confidence limits. These are usually expressed as percentiles; thus, 5 and 95 per cent confidence limits around an estimate indicate a probability of 90 per cent that the actual result lies between these limits and a 10 per cent probability that it lies outside. Unfortunately, for many scenario-based assessments, it is not possible to assign confidence limits to future estimates. However, uncertainties can still be depicted in a relative or qualitative manner. For example, a range of temperature change scenarios has been defined for Finland based on averaged GCM estimates over the region; these are scaled according to IPCC projections of low global GHG and aerosol emissions, combined with a low climate sensitivity (low scenario: 0.1°C warming per decade) and high

BOX 8.2
CROSS IMPACTS ANALYSIS

Cross impacts analysis (Holling, 1978)[23] is a method of highlighting and classifying the relationships between key elements of a system. It entails identifying the pertinent variables of the system and entering these into an interaction matrix, which represents the relationships between the different variables. If one variable exerts a direct influence on the other variable, an entry is made in the appropriate matrix cell.

An example of a cross impacts (structural) analysis for climate change impacts on forests is given in Figure A. Only the presence or absence of an influence is indicated. For example, climate change is shown to influence forest microclimate, fire, tree birth, tree growth, tree death, aquatic ecosystems, wildlife and recreation: a total driving power (row sum) of 8. Similarly, pathogens are influenced by forest

A *An interaction matrix for forest impacts*

Variable Names	Climate change	CO₂ enrichment	Forest microclimate	Fire	Pathogens	Tree birth	Tree growth	Tree death	Forest area	Nutrient cycling	Trace gas emissions	Chemistry	Aquatic ecosystems	Wildlife	Forest products industry	Fisheries	Recreation	Economics	Row sum (driving power)
Climate Change		I	I		I	I	I						I	I			I		8
CO₂ Enrichment	I		I			I	I						I						5
Forest microclimate				I	I	I	I	I		I	I	I	I	I			I		11
Fire					I	I	I	I	I	I	I			I			I		9
Pathogens						I	I	I											3
Tree birth										I				I	I	I			4
Tree growth			I		I					I	I	I		I	I		I		8
Tree death			I		I					I	I	I	I	I	I		I		9
Forest area	I		I							I				I	I		I		6
Nutrient cycling						I	I	I			I		I	I					6
Trace gas emissions												I							1
Chemistry	I					I	I	I											4
Aquatic ecosystems						I											I		2
Wildlife						I	I	I			I	I					I		6
Forest products industry						I	I	I										I	4
Fisheries													I					I	2
Recreation						I	I	I					I	I				I	6
Economics																			0
Column sum (dependency)	3	0	2	5	2	12	10	10	4	5	7	2	7	9	4	1	9	3	

Source: After Martin and Lefebvre, 1993.[24]

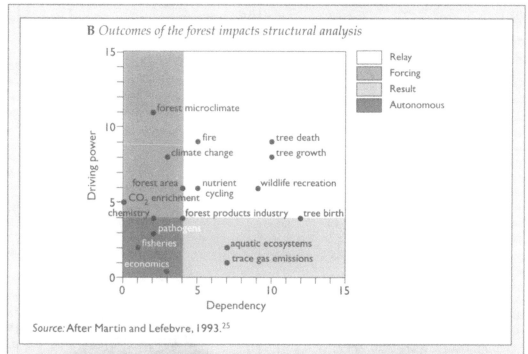

B *Outcomes of the forest impacts structural analysis*

Source: After Martin and Lefebvre, 1993.[25]

microclimate and fire, giving a total dependency (column sum) of 2.

Variables can now be categorized into four types: autonomous variables (weak drivers and weakly dependent), result variables (weak drivers and strongly dependent), relay variables (strong drivers and strongly dependent), and forcing variables (strong drivers and weakly dependent). These can readily be distinguished by plotting the row and column totals on a driving power–dependency graph (Figure B). In this example, forest microclimate appears to be more of a forcing variable than climate change from the standpoint of the forest. Moreover, by these criteria, tree growth and tree death should constitute good indicators of climate change and could be usefully monitored to determine the future of the forest.

emissions combined with a high sensitivity (high scenario: 0.6°C warming per decade).[26] A simple model of crop development was used in conjunction with these scenarios to estimate the potential northward shift in thermal suitability for spring wheat cultivation in Finland by 2050 (see Figure 8.10b). While the confidence in any one scenario cannot be quantified, and while the range depicted does not represent the full range of uncertainty in regional temperature projections, such an analysis does provide an impression of the magnitude of uncertainty in future climate, which can be compared with other sources of uncertainty such as the confidence in the

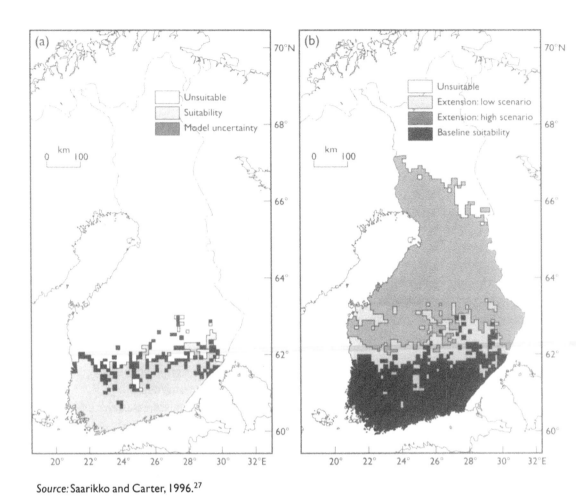

Source: Saarikko and Carter, 1996.[27]

Figure 8.10 *Modelled thermal suitability for spring wheat (var Ruso) in Finland: (a) Uncertainty of the crop development model (95% confidence intervals) under baseline temperatures (1961-90); (b) Uncertainty of the extension in suitability by 2050, attributable to alternative scenarios of temperature change*

crop development model (Figure 8.10a).

CONCLUSIONS FOR STEP 5

Probably the most difficult task facing the assessor of climate impacts is to express the results in a form most valuable to users. Frequently, this requires the generalization of results, often available for a limited number of sites, to the regional and national level at which policy is formulated. It may also involve the interpretation of effects, which may have been estimated as changes in first-order values such as productivity (for example, crop yields or runoff), into forms which affect human livelihood (such as food and water supply). This interpretation requires dealing with uncertainties and may incorporate qualitative description and expert judgement in order to achieve comparable sets of information on the costs and benefits of effects or outcomes.

Chapter 9

THE SIXTH AND SEVENTH STEPS:

Evaluating Adaptive Responses

There are two types of response to climate change and its consequences: mitigation and adaptation. These were described briefly in Chapter 4 and illustrated diagramatically in Figure 4.1. The evaluation of mitigation policies is beyond the scope of this book and is described elsewhere.[1] However, identifying and evaluating adaptation options is an essential component of impact assessment.

A basic distinction is drawn between responses to climate change that are automatic or built-in (termed autonomous adjustments), and responses that require deliberate policy decisions, described as adaptation strategies. While there are some overlaps between these two types of adaptation, they are allocated separate steps in the IPCC assessment framework (see Figure 1.1 in Chapter 1) in recognition of the different treatment they usually receive in assessment studies.

STEP 6: ASSESSMENT OF AUTONOMOUS ADJUSTMENTS

Most ecological, economic or social systems will undergo some natural or spontaneous adjustments in the face of a changing climate. These 'autonomous' adjustments are likely to occur in response both to gradual changes in average climate (which may be barely perceptible relative to background climatic variability) as well as to more drastic shifts in climate – for example, those associated with a change in dominant atmospheric circulation patterns. What is much less certain, however, is what forms these adjustments will take and what costs they will incur. Clearly, in order to obtain credible estimates of impacts, there is a need to account for these autonomous adjustments in the assessment process. It is useful to distinguish three types of autonomous adjustment: inbuilt, routine and tactical adjustments.

Inbuilt adjustments

Inbuilt adjustments, sometimes referred to as physiological adjustments, are the unconscious or automatic reactions of an exposure unit to a climatic perturbation. Some of these are easy to identify (for example, the automatic response of a plant to drought conditions is to reduce transpiration by closing its stomata) and can be accounted for in models that describe the system. Others are more difficult to detect (for instance, the ability of long-lived organisms such as trees to acclimatize to a slowly changing climate). These may require the implementation of some controlled experiments to determine the nature of the adjustment mechanisms (for example, by transplanting tree species between different climatic regimes to investigate the processes of acclimatization in a changed environment).[2]

Routine adjustments

Routine adjustments refer to everyday, conscious responses to climatic variations that are part of the routine operations of a system. For example, as climate alters, the growing season for crop plants also changes, and crop performance might be improved by shifting the sowing date. In some crop-growth models the sowing date is determined by climate (for example, the start of the rainy season or the disappearance of snow cover), so it is selected automatically to suit the conditions. Here, the model is performing internally an adjustment that a farmer might do instinctively or routinely.

Tactical adjustments

Tactical adjustments imply a level of response over and above the adjustments that are made routinely in the face of climatic variability. Such adjustments might become necessary following a sequence of anomalous climatic events, which indicate a shift in the climate. For example, a run of years with below-average rainfall in a semi-arid region may persuade farmers that cultivation of a drought-resistant crop such as sorghum is more reliable than a drought-sensitive crop such as maize, in spite of its lower yield capacity than maize in favourable conditions. Adjustments of this type require a behavioural change, but can still be accommodated internally within the system.

In moving towards a more interventionist type of adjustment, however, the distinction between autonomous adjustments and adaptation starts to become blurred. For instance, it is not always a straightforward task to separate out autonomous tactical adjustments that are directly related to climate change from adjustments that are made to changing external conditions, which are themselves an adaptive response to climate change (such as government assistance to farmers to cope with adverse climatic conditions). The evaluation of these exogenous adaptations is examined later in this chapter.

INCORPORATING ADJUSTMENTS IN INTEGRATED MODELS

Sectorally integrated models

There are increasing efforts to incorporate autonomous adjustments into impact models. As an example, consider the effect of climate change on altered yield potentials of major food crops (discussed in Chapter 8). These effects will almost certainly be buffered by changes in the land area farmed, which will itself respond to changes in price. It is therefore necessary to model the yield changes in the context of supply, demand, the potential unfarmed area, and commodity prices that are affected by these. Such an assessment was conducted using a world food model which simulated how different countries, experiencing different changes in yield or output potential, might fare in the world food system (this was described in Chapter 8).[3] The resulting estimates for world cereal production are shown in Figure 9.1a. These results assume, unrealistically, no farm-level adjustment or adaptation to climate change. The next step was to include minor farm-level autonomous adjustments, such as changes in planting date, in amounts of irrigation and in choice of crop varieties that are currently available (Figure 9.1b), and then major adjustments, such as expansion of irrigation systems and development of new cultivars (Figure 9.1c). The conclusion was that only major adjustments led to significant reductions of impact.

Intersectorally integrated models

More comprehensive integration of adjustments in climate impact modelling has been achieved through the simulation of interactions between sectors (such as altered prices and demand) and of multiple effects on related sectors (such as effects of sea-level rise, altered runoff affecting water available for irrigation, as well as changes in weather). An example is the integrated assessment of impacts on agriculture in Egypt, which incorporated altered world food prices (from the international food study described above),[4] land lost to sea-level rise and altered levels of Nile water which affect irrigation.[5] A summary of this study is given in Box 9.1.

STEP 7: EVALUATING ADAPATION STRATEGIES

Under many of the climate changes anticipated for the future, a wider array of adaptations, in addition to those autonomous adjustments described above, are likely to be needed in order to reduce impacts significantly. As distinct from being autonomous these may require deliberate policy decisions. The following section considers a general means of evaluating such adaptations. It comprises seven broad steps (see Figure 9.2) which are treated, in turn, below.[6]

1 Defining the objectives

Any analysis of adaptation must be

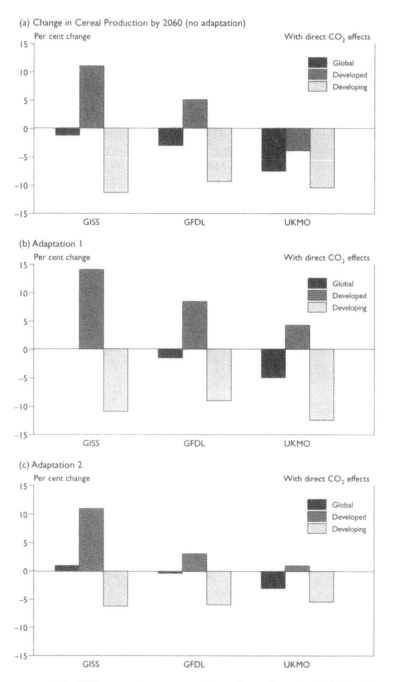

Note: Reference scenario for 2060 assumes future output with no climate change (world: 3286 million metric tons production; developed countries: 1449 mmt; developing countries: 1836 mmt). GISS, GFDL and UKMO are GCM-based climatic scenarios (see Box 9.1). For explanation, see text.

Source: Rosenzweig and Parry, 1994.[7]

Figure 9.1 *Efficacy of adaptation options employed in the world food study*

BOX 9.1

CASE STUDY: AN INTEGRATED ASSESSMENT OF IMPACTS OF CLIMATE CHANGE ON THE AGRICULTURAL ECONOMY IN EGYPT

Background: agriculture in Egypt is restricted to the fertile lands of the narrow Nile Valley from Aswan to Cairo and the flat Nile Delta north of Cairo. Together this comprises only 3 per cent of the country's land area. Egypt's entire agricultural water supply comes from irrigation, solely from the Nile River. In 1990, agriculture (crops and livestock) accounted for 17 per cent of Egypt's gross domestic product.

Problem: the study sought to assess the potential impact of a change in climate and sea level on Egypt's agricultural sector, accounting for changes in land area, water resources, crop production and world agricultural trade. The aim was not to predict Egypt's future under a changed climate, but rather to examine the combined effects on agriculture of different natural factors and the adaptability of the economic system.

Methods: the assessment was part of an international study of climate change impacts on world food supply and trade,[8] forming one component of a coordinated international programme of climate change impact studies.[9] A number of submodels were used to estimate the different sectoral impacts of climate change

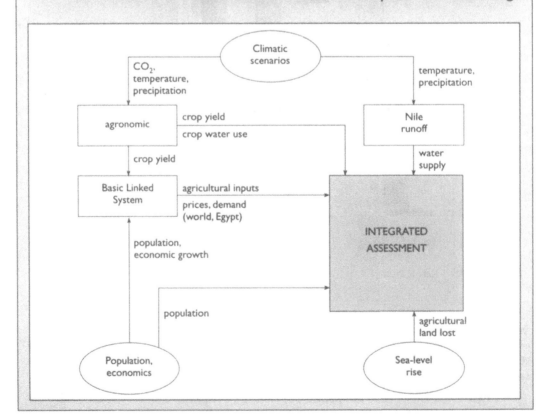

(see figure). A digital elevation model of the Nile Delta was developed for determining land loss due to sea-level rise. A physically based water-balance model of the Nile Basin was used to evaluate river runoff. This was linked to a simulation model of the High Aswan dam complex to determine impacts on Lake Nasser yields. Process-based agronomic models (incorporating direct effects of elevated CO_2) were used to estimate crop yields and crop water requirements, and cropping patterns under different climatic scenarios were determined using the Egyptian food supply and trade model: one component of an international food trade model, the Basic Linked System (BLS), which was run at a global level. Results from the BLS and other submodels were then taken directly, or aggregated using expert judgement, to provide inputs to an Egyptian Agricultural Sector Model (EASM-CC). This is an integrated model of the agricultural economy incorporating effects on water, land, crops, livestock and labour. One output of the model is the annual consumer–producer surplus, an economic measure of social welfare.

Testing of methods: each of the submodels used in the study was validated against local data. Furthermore, an elaborate comparative analysis was undertaken to select an appropriate hydrological model from a number of candidate models. Each of the linked national or regional models in the BLS has been tested in its region of origin, while the complete model was initialized with 1980 data from the Food and Agriculture Organization (FAO) and run through to 1990, model parameters being tuned for the 1980s period to obtain the 'best fit' for 1990.

Scenarios: the current baseline adopted for the socio-economic projections was 1990 and the climatological baseline, 1951–80. The time horizon of the study, 1990–2060, was largely dictated by the climate change projections. Socio-economic scenarios for a future world in 2060 were developed for population (estimated from UN/World Bank projections to more than double, assuming current growth rates) and economic growth (based upon growth rates assumed in the world food supply and trade study).

The climatic scenarios were based on three equilibrium 2 x CO_2 GCM experiments (each displaying results close to the upper end of the 1.5–4.5°C range of global mean annual temperature projections given by the IPCC) and a fourth 'low-end' scenario (in the middle of this range), based on transient model outputs. Each scenario comprised values of mean monthly changes in temperature, precipitation and solar radiation. Values from the appropriate GCM grid box were applied as adjustments to local daily or monthly climatological observations for the baseline period. The scenarios were assumed to apply in 2060, and to coincide with a CO_2 level of 555 ppmv, broadly similar to the IPCC IS92a projection (see Box 7.2).

Sea-level rise associated with changing temperatures was estimated to be 37 centimetres between 1990 and 2060. This estimate is derived from a one-metre global sea-level rise by 2100, the same scenario as that used in the IPCC Common Methodology (see Box 9.2) but at the high end of recent estimates (see Table 7.1). This was added to a predicted 38 centimetre subsidence of the Nile Delta, giving a relative sea-level rise of 75 centimetres by 2060.

Impacts: impacts were estimated as the difference between simulations for 2060 without climate change, based on projections of population, economic growth, agricultural production, commodity demand, land and water resources and water use (base 2060), and simulations with changed climate according to the four climatic scenarios.

The table below provides a summary of the impacts of the four scenario climates on each sector, together with the integrated impacts on economic welfare (the consumer–producer surplus). The agricultural water productivity index is an aggregate measure of impacts on agriculture: total agricultural production (tonnes) divided by total agricultural water use (cubic metres). The results illustrate how impacts on individual sectors are affected by impacts on other sectors. For example, under the GISS scenario, despite an 18 per cent increase in water resources, the 5 per cent loss of land and 13 per cent reduction in agricultural water productivity leads to a 6 per cent reduction in economic welfare. The results also demonstrate how individual sectoral assessments may give a misleading view of the overall impact, which is better reflected in the integrated analysis. For instance, under the 'low-end' scenario, while sectoral impacts are mainly positive, the integrated impact is actually a 10 per cent decline in economic welfare. This is because the rest of the world performs better than Egypt under this scenario; Egypt loses some of its competitive advantage for exports and thus the trade balance declines.

Adaptive responses: adaptations in water resources (major river diversion schemes), irrigation (improved water delivery systems), agriculture (altered crop varieties and crop management) and coastal protection against sea-level rise were all tested for the UKMO scenario. They achieve a modest 7 to 8 per cent increase in agricultural sector performance compared to no adaptation, but together would be extremely expensive to implement. However, investment in improving irrigation efficiency appears to be a robust, 'no regrets' policy that would be beneficial whether or not the climate changes.

Source: Strzepek et al, 1995.[10]

A comparison of sectoral with integrated impacts for the four climatic scenarios (per cent change from 2060 base results).

| Climatic scenario | Land area | Sectoral impacts | | | Integrated impact |
		Food demand	Agricultural water productivity index	Water resources	Consumer–producer surplus
UKMO*	−5	−3	−45	−13	−23
GISS**	−5	−1	−13	+18	−6
GFDL***	−5	−1	−36	−78	−52
'Low-end'	−5	0	+10	+14	−10

*   United Kingdom Meteorological Office model[11]
**   Goddard Institute for Space Studies model[12]
*** Geophysical Fluid Dynamics Laboratory model[13]

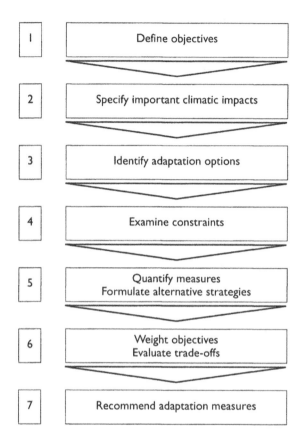

Source: Carter et al, 1994.[14]

Figure 9.2 *Development of an adaptation strategy*

guided by some agreed goals and evaluation principles. It is not sufficient to state that adverse impacts should be avoided, reduced or eliminated. Two examples of general goals commonly propounded by international institutions and conventions are: the promotion of sustainable development, and the reduction of vulnerability. However, these are so broad that they are open to many different interpretations; therefore, specific objectives need to be defined that complement the goals.

Objectives are usually derived either from public involvement, from stated public preferences, by legislation, through an interpretation of goals such as those described above, or any combination of these. They represent desired targets which can be evaluated using specified criteria and constraints. Table 9.1 illustrates three possible objectives that might be selected to achieve each of the two different goals described above, and the

Table 9.1 *An example of multiple criteria evaluation for water resources management*

Overall Goal	Specific Objective	Evaluation Criteria
Sustainable development	1 Regional economic development	Income
		Employment
	2 Environmental protection	Biodiversity
		Habitat areas
		Wetland types
	3 Equity	Distribution of employment
		Minority opportunities
Reduce vulnerability	1 Minimize risk	Population at risk
		Frequency of event
	2 Minimize economic losses	Personal losses
		Insured losses
		Public losses
	3 Increase institutional response	Warning time
		Evacuation time

Source: Stakhiv, 1994.[15]

evaluation criteria that can be used to measure their success.[16] Most of these are quantitative measures (such as income and employment); others such as biodiversity can be quantified, but not usually in economic terms.

2 Specifying the important climatic impacts

This step involves an assessment, following the methods outlined elsewhere in this book, of the possible impacts of climatic variability or change on the exposure unit. Where climatic events are expected that will cause damage, these need to be specified in detail so that the most appropriate adaptation options can be identified. Where beneficial climatic events are anticipated, these should be examined, both in their own right and because they may help to compensate for negative effects. Details include the magnitude and regional extent of an event, its frequency, duration, speed of onset and seasonality (timing during the year). In the case of long-term climate change, impacts should be considered relative to those that would be expected to occur in the absence of climate change (the future baseline; see Chapter 7). Moreover, since there is often great scientific uncertainty attached to projections, it may be useful to express possible changes in terms of the probability of their occurrence or as changes in the recurrence frequency of events observed in the historical climatological record.

One general approach for identifying the exposure units at risk from climate variability is vulnerability assessment. Vulnerability can be defined as the degree to which an exposure unit is disrupted or adversely affected as a result of climatic events. It follows that vulnerable systems, activities or regions are likely to be those most in need of planned adaptation. The approach can be illustrated with reference to a common methodology that has been developed for the national-scale assessment of coastal zone vulnerability to sea-level rise (see Box 9.2).[17] One of the main objectives of the common methodology is to inform national decision-makers about the vulnerability of the coastal zone, the possible problems a country may face due to a changing climate and sea level and, if necessary, the types of assistance that are most needed to overcome these problems.

3 Identifying the adaptation options

Six generic types of behavioural adaptation strategy for coping with the negative effects of climate have been identified by Burton and colleagues:[18]

- *prevention of loss*, involving anticipatory actions to reduce the susceptibility of an exposure unit to the impacts of climate (eg controlled coastal zone retreat to protect wetland ecosystems from sea-level rise and its related impacts);
- *tolerating loss*, where adverse impacts are accepted in the short

term because they can be absorbed by the exposure unit without long-term damage (eg a crop mix designed to minimize the maximum loss, to ensure a guaranteed minimum return under the most adverse conditions);
- *spreading or sharing loss*, where actions distribute the burden of impact over a larger region or population beyond those directly affected by the climatic event (eg government disaster relief);
- *changing use or activity*, involving a switch of activity or resource use from one that is no longer viable, following a climatic perturbation, to another that is, in order to preserve a community (eg by employment in public relief works);
- *changing location*, where preservation of an activity is considered more important than its location, and migration occurs to areas that are more suitable under the changed climate (eg the resiting of a hydro-electric power utility due to a change in water availability);
- *restoration*, which aims to restore a system to its original condition following damage or modification due to climate (eg an historical monument susceptible to flood damage); this is not strictly an adaptation to climate, as the system remains susceptible to subsequent comparable climatic events.

Box 9.2

CASE STUDY: EFFECTS OF CLIMATE CHANGE ON COASTAL
ENVIRONMENTS IN THE MARSHALL ISLANDS

Problem: for many low-lying coastal areas of the world, the effects of accelerated sea-level rise (ASLR) associated with global climate change may result in catastrophic impacts in the absence of adaptive response strategies. Even in the absence of climate change, however, the combined pressures of growth and development will require organized adaptive response strategies to cope with an increased vulnerability of populations and economies to storms, storm surges and erosion. The Republic of the Marshall Islands consists of 34 atolls and islands in the Pacific Ocean with majority elevations below two to three metres above mean sea level. A vulnerability analysis case study for Majuro Atoll was conducted to provide a first-order assessment of the potential consequences of ASLR during the next century.

Method: the study followed a common methodology outlined by the IPCC.[19] That methodology follows, in some respects, the general framework identified by the seven steps described in this book. The study was concerned only with the effects of ASLR (inundation, flooding, groundwater supplies), leaving the integration of frequency and intensity of extreme events, changes in currents and tides, increased temperature and changes in rainfall patterns for the future, when regional models can simulate such changes.

Testing of method: the study included a multidisciplinary team made up of in-country experts, regional assistance from the South Pacific Regional Environment Programme and a consulting firm which conducted oceanographic–engineering studies. The methodology proved very useful in identifying potential impacts to atolls and adaptation responses. Reliance on existing information and lack of other information placed some limitations on the study, but qualitative data obtained during the study permitted meaningful extrapolations.

Scenarios: ASLR of one metre by the year 2100 was used to assess the worst case impact to shoreline communities. Three scenario cases were considered (as specified by the common methodology):

1 ASLR=0 for zero sea-level rise;
2 ASLR=1 for a 0.3m (1.0 ft) rise; and
3 ASLR=3.3 for a 1 m (3.3 ft) rise.

Subsidence/uplift or regional variability were not taken into account due to lack of information. The effects were considered for both the ocean and lagoon side of the atoll and for four major study areas representing most environmental conditions of the atoll nation.

Impacts: the potential effects of ASLR include:

1. an approximate 10 to 30 per cent shoreline retreat with a dry land loss of 160 acres out of 500 acres on the most densely populated part of the atoll;
2. a significant increase in severe flooding by wave runup and overtopping with ASLR=3.3, resulting in flooding of half of the atoll from even normal yearly runup events;
3. flood frequency increases dramatically;
4. a reduction of the freshwater lens area which is important during drought periods;
5. a potential cost of protecting a relatively small portion of the Marshall Islands of more than four times the current GDP;
6. a loss of arable land, resulting in increased reliance on imported foods.

Policy options: the study considered, though did not formally evaluate, the options of protection (including structural considerations), accommodation (including land elevation and adaptive economic activities for flooded areas), a retreat strategy to the highest elevations on the atoll, and a no-response strategy (including a continuation of ad hoc and crisis response measures currently used to address flooding problems). The major recommendations included the need to develop and implement integrated coastal zone management, which would incorporate ASLR response planning and begin the process of understanding the natural and human systems likely to be affected by climate change.

Source: Hotthus et al, 1992.[20]

A complete list of strategies should be developed from a survey of past experience, employing the following kinds of questions:[21]

- How much warning of the events was available?
- Was the warning given to those likely to be impacted?
- Did those who received the warning act upon it?
- Did they know what to do and were they advised or assisted in taking action?
- Could forecasting and warnings and their dissemination be improved?
- What longer-term anticipatory actions were taken to prevent such events or to reduce their impacts?
- How much damage was caused?
- Can the damage or impact data be broken down into categories, such as fatalities, injuries, ill-health, malnutrition, physical property, damage to crops, forests, infrastructure, housing, public buildings, communications and transport facilities?
- Looking over recent experience, are there trends in levels of

damage?
- What steps were taken before, during and after the events to reduce their impact?
- What action was taken at various levels of decision-making?
- What action was taken by governments at the national, regional or local scale?
- What actions were taken by individuals at risk and by the private sector?
- How successful were these response actions?
- How might they be performed differently with respect to future events?
- What obstacles were encountered in attempting to respond to the events?
- How could these obstacles be removed, or incentives created to improve future response?
- What sort of loss-sharing took place at family, community, regional and national levels?
- Was the international community involved in providing assistance?

Different criteria can be used for organizing the information. For instance, detailed tables have been used to catalogue traditional adjustment mechanisms for coping with interannual climatic variability in self-provisioning societies.[22,23] Table 9.2 illustrates a classification system for displaying smallholder coping strategies for drought in central and eastern Kenya, as well as a qualitative effectiveness ranking for different measures.

Other methods of cross-tabulation have been employed in formal procedures of resource management. For example, alternative water-resource adaptation measures in the United States are commonly analysed according to both the type of measure and its strategic scope. Four groupings of strategy have been identified:[24]

- long-range strategies, generally pertinent to issues involving mean changes in climate (eg river basin planning, institutional changes for water allocation);
- tactical strategies concerned with mid-term considerations of climatic variability (eg flood proofing, water conservation measures);
- contingency strategies, relating to short-term extremes associated with climatic variability (eg emergency drought management, flood forecasting);
- analytical strategies, embracing climatic effects at all scales (eg data acquisition, water management modelling).

Numerous options exist for classifying adaptive measures, but generally – regardless of the resource of interest (such as forestry, wetlands, agriculture, water) – the prospective list should include management measures which reflect:

Table 9.2 *Characteristics of selected coping strategies by smallholders for drought in central and eastern Kenya[a]*

Response/coping strategy	Prevalence	Effectiveness			Recovery	Constraints				
		Normal	Moderate drought	Severe drought		Labour	Capital	Technology	Education/information	Land
Subsistence production										
Soil conservation	M–H	M	M	L	0	–	–	–	+	–
Water conservation	L	M	H	L	0	–	–	–	+	–
Irrigation	L	H	H	H?	+	–	+	+	+	+
Multiple farms	L	M	M	L	0	–	+	–	–	+
Inter/relay cropping	H	H	H–L	L	+	–	–	+	+	–
Dry planting	M	H	H	L	0	–	–	–	–	–
Mixed livestock herds	M–H	H	H	M	+	+	–	–	–	+
Dispersed grazing	H	H	H	M	+	+	–	–	–	+
Fodder production[b]	M	H	H	M	+	+	–	+	+	+
Drought-resistant crops	H	M	H	M	+	–	–	+	+	–
Monetary activity										
Local wage labour	M–H	H	H	H	+	+	–	–	+	–
Migrant wage labour	M	H	H	H	–	+	–	–	+	–
Permanent employment	M	H	H	H	+	+	–	+	+	+
Local business	L	M	M	L	0	+	+	–	+	–
Cash crop[c]	M	H	M	L	+	+	+	+	+	+
Sell capital assets	M	H	M	L	–	–	–	–	+	+
Livestock sales	H	H	M–H	M–L	–	–	–	–	+	–
Remittances/donations										
Relatives/friends	M–H	M	H	H	+	–	–	–	–	–
Government and others[d]	M–H	?	H	H	?	–	+	–	+	–
Loans/credit	L	H	H	H	+	–	+	–	+	–

Key:

Prevalence/Effectiveness
H   = >50%/High
M–H = 30–50/Mod.high
M   = 15–30/Moderate
L   = 0–15/Low

Recovery/Constraints
– = negative, ie impedes recovery/is a constraint
+ = positive, ie aids recovery/is not a constraint
0 = neutral, ie no effect on recovery
? = uncertain or variable

a Consensus agreement by authors based on available data. Ratings are intended to be qualitative and relative as no systematic survey data are available. In many cases, these are hypotheses to be verified. *b* Very common in the upper altitudinal zones; almost non-existent in the lower zones. *c* Does not include food crops. Very common in upper zones; rare in lower zones. *d* High in the lower zones; low in the upper zones.

Source: Adapted from Akong'a et al, 1988.[25]

- structural/infrastructural measures;
- legal/legislative changes;
- institutional/administrative/ organizational measures;
- regulatory measures;
- education;
- financial incentives, subsidies;
- research and development;
- taxes, tariffs, user fees;
- market mechanisms;
- technological changes.

4 Examining the constraints

Many of the adaptation options identified in the previous step are likely to be subject to legislation, influenced by prevailing social norms related to religion or custom, or constrained physically (such as the landward retreat from an eroding coast) or biologically (for example, genetic material for plant breeding to adapt to a changing climate). This may encourage, restrict or totally prohibit their use. Thus, it is important to examine closely, possibly in a separate study, what these constraints are and how they might affect the range of feasible choices available. As stated by Burton in the draft version of the UNEP *Handbook on Climate Impact Assessment*, five questions relate to the constraint on feasibility:[26]

- What is technically possible? Technology offers a wide range of possibilities for managing human activities in ways that make them less vulnerable to

climatic events. In addition, new technological possibilities are continuously being created. A technology assessment could therefore form a useful part of an adaptation study.
- What is economically gainful? Not all technology and not all adaptation options are efficient in an economic sense. It is therefore essential to examine adaptation options in terms of their expected benefits (damages prevented or reduced) against their cost.
- What is socially or legally acceptable? Not all adaptation options conform to socially accepted behaviour or norms, and some may be illegal. It is therefore advisable to examine proposed adaptation options against such social and legal constraints. This does not mean that such constraints are necessarily valid. Statutes, customs and traditional ways of doing things can be an obstacle to effective adaptation.
- What is environmentally friendly or sustainable? The possibility has to be considered that potentially useful adaptation options might have adverse consequences for the environment, and if there are significant reasons to suppose that this might be so, an environmental impact assessment could be undertaken.

• What regional or spatial links are involved? Adaptation options that promise to reduce vulnerability in one place or region can have adverse effects elsewhere. In some cases, therefore, a regional or spatial analysis of the effects of adaptation options may be required. This can apply to regions within a country, and also to transboundary effects on neighbouring countries.

5 Quantifying the measures and formulating alternative strategies

Assessing the effectiveness of management measures

Once management measures have been identified, their performance needs to be assessed with respect to the stated objectives. In some cases, when appropriate data and tools are available, simulation models can be used (for instance, by simulating the effectiveness of different measures under a range of climatic scenarios). In other cases, published evidence, survey material or expert judgement can be used. Both methods yield either quantitative or qualitative results. In this way, it may be possible to obtain a relative, or in some cases an absolute, evaluation of the comparative effectiveness of different measures in fulfilling individual objectives. A formal and replicable method of conducting such an assessment is *multicriteria analysis*.

It is at this stage that considera-

tions of costs and benefits are likely to play a prominent role in the evaluation. Moreover, the desired rate of adaptation is likely to have a strong influence on the economic viability of certain measures. The faster the need for adaptation, the more it is likely to cost per annum. A limited capability now exists for testing the feasibility of different kinds of adaptation by altering input data or assumptions in simulation models. A detailed example of this is given in Box 9.3.[27]

Formulation of a multiobjective strategy
The above analysis can then be extended to assess the effectiveness of each measure across all the objectives, recognizing that some objectives may conflict with each other (regional economic development often conflicts with environmental protection). This is a prelude to the development of strategies. Each strategy involves a set of management measures that maximize the level of achievement of some objectives, without jeopardizing progress towards the remaining objectives.

Step 5 is illustrated schematically in Figure 9.3. The top half portrays a multicriteria analysis. Each row represents a different management measure (mm) and each column a different objective. The objectives satisfy a given goal, and the effectiveness of a management measure in achieving an objective is assessed according to evaluation criteria (see Table 9.1). The tinted areas show the highest ranked (most effective) measures for satisfying the objective in column 1, and the

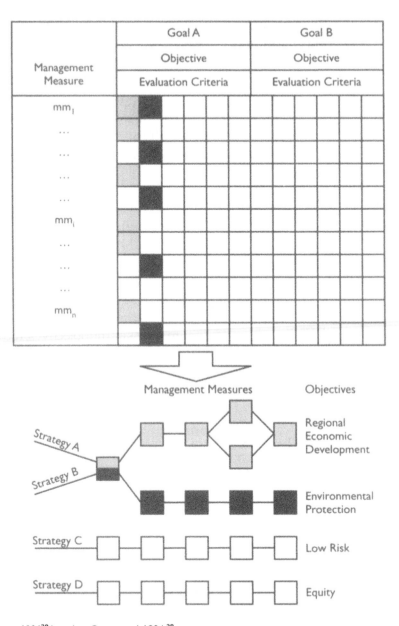

Source: Carter, 1996[28] based on Carter et al, 1994.[29]

Figure 9.3 *Some procedures for formulating adaptation strategies (top: multicriteria analysis; bottom: multiobjective strategy formulation)*

black shaded areas show the highest ranked measures for satisfying the objective in column 2. The bottom half of Figure 9.3 shows how these management measures might be organized into alternative strategies (multiobjective strategy formulation). This example is simplistic; each strategy needs to consider the effectiveness of the measures across all objectives, rather than independently. Moreover, the strategy formulation also needs to account for the sequence (timing) and discounted cost of such measures.

6 Weighting objectives and evaluating trade-offs

Steps 1 to 5 in the assessment frequently yield conflicting results (such as adaptation strategies relating to specific objectives – see Figure 9.3, bottom). Hence, Step 6 of the analysis is a key part of the evaluation: it seeks to resolve the conflicts by attaching weights to different objectives according to assigned preferences and then comparing the effectiveness of different strategies in meeting these revised objectives. For instance, one coastal response to accelerated sea-level rise might be a 'pure retreat' option, consisting exclusively of regulatory measures, taxes, incentives and legal and institutional measures spread over a period of time. Alternatively, a 'pure protection' option might be considered, involving structural measures and better organized monitoring, warning and evacuation

plans. The outcomes of these responses would be quite different, so in choosing between them it would be necessary to weigh carefully the objectives and preferences of different stakeholders.

All the component measures of a strategy are compared against the same set of objectives, so that managers, policy-makers and the public can see the relative range of costs and benefits for each strategy, as well as the distribution of impacts among the sectors and population (equity). Only then can compromises or trade-offs be made among objectives and between management measures. A wide variety of methods and models are available for such multicriteria analysis.[30] For example, four basic objectives are used operationally to evaluate the effectiveness of different planning strategies in federal water projects in the United States:[31]

- national economic development (based on measures such as gross domestic product, per capita income, or employment);
- environmental quality (significant environmental resources and their ecological, cultural and aesthetic attributes);
- regional economic development (distribution of regional economic activity in terms of income and employment);
- other social effects (including urban and community impacts, life, health and safety factors, displacement, long-term

> # BOX 9.3
> ## CASE STUDY: THE MINK PROJECT – AN INTEGRATED REGIONAL ASSESSMENT
>
> *Background:* Missouri, Iowa, Nebraska and Kansas (the MINK region) are four adjacent states in the central United States which are dependent on resource sectors known to be sensitive to climate change: agriculture, forestry, water resources and energy. Except for pockets of forestry on the Ozark Plateau of south-east Missouri, and grassland on sand dunes in north-central Nebraska, the region is fairly coherent, with flat or rolling topography used predominantly for agriculture.
>
> *Problem:* to study how climate change might affect the current and future functioning of regional-scale economies.
>
> *Method:* a number of models were used to evaluate impacts of climate on individual sectors. For agriculture a semi-empirical process model (EPIC) was adopted that simulates crop biomass and yield production, evapotranspiration and irrigation requirements. For forests, a succession model (FORENA) that simulates the annual development of individual trees within a mixed-species forest was used. This allows the effects of climate change on both forest growth and species composition to be evaluated. Both EPIC and FORENA were modified to account for the direct effects of CO_2 on photosynthesis and water use. Several approaches were used to estimate regional water resources: changes in evapotranspiration and irrigation requirements were modelled using EPIC. Regional water supply was estimated using empirical relationships between present and past streamflows. Impacts on the energy sector drew on the modelling and interpretations from the other three sectors and on an analysis of how heating and air conditioning requirements are affected by changing temperature. Finally, the economy-wide effects of changes in productivity of the above resource sectors were studied using IMPLAN, a regional input–output model. IMPLAN describes the interaction between 528 industries in the MINK region.
>
> *Testing of methods/sensitivity:* EPIC was validated against national agricultural statistics (county level) and observed seasonal yields in agronomic experiments for seven crops in the region. Evapotranspiration terms were compared with field observations. FORENA had been validated previously for conditions in the eastern United States, and results were also compared with observed forest behaviour under drought conditions in Missouri. A sensitivity study was conducted on the response of forest biomass to changes in temperature and precipitation. The model coefficients relating input and output flows between industries in IMPLAN were computed from regional data for 1982.

Scenarios: a temporal analogue was employed as the climate scenario, specifically the decade 1931–40 in the MINK region. Overall, this period was one of severe drought – both drier and warmer than average in the region, consistent in sign with GCM projections (see Figure 7.2). These conditions were assumed to occur in the present and also in the year 2030, along with an increase in CO_2 concentration of 100 ppm (to 450 ppm). Four sets of conditions were investigated:

1   the current baseline, which referred to the economic situation in the early 1980s, with 1951–80 as the climatological baseline;
2   climate change imposed on the current baseline;
3   a baseline description of the economic structure of the region in the future without climate change (including population, economic activity and personal income); and
4   imposition of climate change on the future baseline (including feedbacks between sectors, such as the projected extent of irrigated agriculture given scenarios of future water supply).

Impacts: in the MINK region of 2030 with a climate like that of the 1930s, the main results of the study are:

1   Crop production would decrease in all crops except wheat and alfalfa, even accounting for CO_2 effects. However, impacts on agriculture overall would be small given adaptation, though at the margins losses could be considerable (eg a shift in irrigation from west to east).
2   Impacts on forestry would not be fully felt by 2030, but in the long term they would be severe. There is little potential for adaptation to the climate change unless the production of wood for biomass fuel makes adaptation economically justifiable. However, forestry is a very small part of the MINK economy.
3   Impacts on water resources would be major and severe. The quality and quantity of surface waters would diminish, water-based recreation would suffer losses and navigation would become uneconomic on the Missouri. Rising costs of water extraction would make agriculture less competitive for surface water and groundwater supplies and would hasten the abandonment of irrigation in the western portions of the region.
4   Only about 20 to 25 per cent of the region's current energy use would be sensitive to a 1930s-type climate, and any impacts of climate would be eased by adjustments within the energy sector so that the effect on the regional economy would be minor.
5   Unless the climate-induced decline in feedgrain production falls entirely on animal producers in MINK (which would lead to an overall loss to the regional economy of 10 per cent), the regional economic impacts of the climate change would be small. This is because agriculture, while important

relative to other regions of the US, is still only a small (and diminishing) part of the MINK economy.

Adaptation: most of the work on adaptation dealt with responses to impacts on crop production. Simulated adjustments included changed planting dates, altered varieties and changed tillage practices. In addition, technological advances were assumed in irrigation efficiency and crop drought resistance as well as improvements in a number of crop specific characteristics, including harvest index, photosynthetic efficiency and pest management. In economic terms, in the absence of onfarm adjustments and CO_2 enrichment, the analogue climate would reduce the value of 1984–87 crop production in MINK by 17 per cent. The CO_2 effect would reduce the loss to 8 per cent, and onfarm adjustments would reduce it further to 3 per cent. In the forestry sector a number of management options were investigated using the FORENA model (eg the suitability of pine plantations and various thinning strategies), but none was considered appropriate as a response to climate change in the region. Rather, barring major public intervention, only reactive measures to forest decline such as salvage cutting were judged likely. Qualitative assessments were made of possible adjustments in water use (eg water conservation, a shift of emphasis from navigation to hydropower production, recreation and water supply on the Missouri River) and in energy production and use (eg energy-saving water pumping and irrigation practices, improved energy-use efficiency and adoption of new or existing technologies for improving electric conversion efficiency and reducing cooling water requirements).

Policy options: although the MINK study did not seek to provide specific recommendations to policy-makers on how to cope with climate change in this region, one important conclusion was that the relevant policy issue at the regional scale is not one of climate change abatement (which can only be dealt with at national and international level), but rather one of optimal adjustment to climate change. An important assumption of the study was that markets play a major role in inducing adjustments needed for adequate response to climate change. However, some important elements are not commonly considered in economic terms (eg the quality of aquatic habitats is projected to decline in the MINK region). These should necessarily fall within the ambit of public policy. Moreover, the study also speculated on possible policy shifts, which could have more far-reaching implications for the MINK region (eg the removal of subsidies for irrigation agriculture under sharply increased water scarcity or subsidization of plantation forestry as a method of capturing atmospheric carbon and of energy production).

Source: Rosenberg, 1993.[32]

productivity, energy requirements and energy conservation).

All impacts and adaptation measures are evaluated according to these four categories. Selecting preferred strategies thus requires determining the trade-offs among the categories. One possible method of evaluating these trade-offs is the policy exercise. Policy exercises combine elements of a modelling approach with expert judgement, and were originally advocated as a means of improving the interaction between scientists and policy-makers. Senior figures in government, industry and finance are encouraged to participate with senior scientists in exercises (often based on the principles of gaming), where they are asked to judge appropriate policy responses to a number of given climatic scenarios. Their decisions are then evaluated using impact models.[33] The method has been tested in a number of climate impact assessments in south-east Asia.[34]

7 Recommending adaptation measures

The results of the evaluation process should be compiled in a form that provides policy advisors and decision-makers with information on the best available adaptation strategies. This information should indicate some of the assumptions and uncertainties involved in the evaluation, as well as the rationale used (such as decision rules, weightings, key constraints, institutional feasibility, national and international support, technical feasibility).

CONCLUSIONS FOR STEPS 6 AND 7

Relatively little analysis has been made of the efficacy of different adaptive strategies at reducing impacts from climate change (as distinct from mitigative strategies that reduce greenhouse gas emissions). One reason for this is that there are very many different types of adaptation that could be adopted and thus need to be evaluated. Some, such as the construction of reservoirs, are technical responses to improve supply and can be readily costed. But the management of demand for water through fiscal, regulatory or other schemes is less easy to design, and most studies completed so far have concluded that the likely optimal path is a combination of technological, economic and social measures to alter both demand for, and supply of, climatic resources such as water.

Furthermore, the separation of impact assessment and adaptation assessment is an artificial contrivance: common sense suggests that some adaptation is bound to take place at the local level. For instance, when we seek to evaluate the effect of climate change on food supply in the future, we should assume that new varieties of crops and improved methods of their management will be available.

It is the additional innovations that we need to evaluate so that we can, ultimately, form a judgement as to the best combination of adaptation to impacts and mitigation of emissions that we should seek.

Chapter 10

CONCLUSIONS:

Organizing the Research and Communicating Results

This book illustrates and expands upon the Technical Guidelines for Assessing Climate Change Impacts and Adaptations, published in 1994 by the Intergovernmental Panel on Climate Change.[1] It includes a number of examples of recent assessments that relate to each of the seven steps advocated in the IPCC approach. It has been emphasized that the seven-step method is not intended to be prescriptive. There is no single 'best' approach. Nonetheless, most of the basic tools and methods used in climate impact assessment are embraced in the IPCC approach.

WORKING THROUGH THE SEVEN-STEP METHOD

In practice, few studies proceed monotonically through all of the steps and many repeat a number of iterations. For example, where the outputs of one impact model are used as inputs to another, similar procedures

for, inter alia, data acquisition, model testing, fixing of assumptions and scenario development should occur when applying each model. In general, we can identify four stages of iteration through which an assessment may need to proceed (see Figure 10.1):

1 feasibility;
2 assessment of biophysical impacts;
3 assessment of socio-economic impacts;
4 évaluation of adaptation options.

These stages are depicted as rows in Figure 10.1. The columns represent broadly comparable steps, with some of the alternative procedures at each step listed in the boxes. Thick arrows show the linkages between stages and thin arrows link the steps. Broken lines represent iterations that may be required to repeat an analysis under a new set of assumptions.

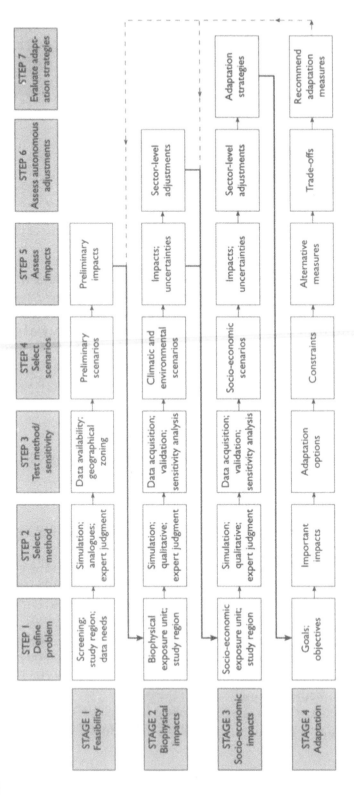

Figure 10.1 *A four-stage method for conducting climate change impact and adaptation assessments*

ENSURING COMPATIBILITY WITH OTHER ASSESSMENTS

Policy-makers may frequently need to take into account a number of trends (in addition to climate change) which will affect the future for which they are planning. For example, while climate change may alter the potential for natural regeneration of mangrove forests, these forests are now experiencing a major reduction in some parts of the world due to land reclamation for agriculture and shrimp farming. Likewise, while changes in rainfall may alter the rate of recharge of a watertable used for irrigation, this will also be affected by the rate of current and future abstraction. In both cases, and in many others where the resource base is affected both by climate change and other physical or socio-economic factors, comparable assessments need to be made. To achieve this they need to adopt, as far as possible, similar assumptions and to develop modes of research that allow for the best possible interconnection. Standard guides in methods of environmental impact assessment are now available and have been adopted by the planning community in North America and Europe.[2] These should be consulted when developing a project on climate impact assessment which specifically needs to take non-climatic trends and factors into account.

SUMMARY AND MAIN FEATURES OF INTEGRATED ASSESSMENT MODELS

Short title	Full name of model and reference	Main features						
		A	*B*	*C*	*D*	*E*	*F*	*G*
Policy Evaluation Models								
AIM*	Asian–Pacific Integrated Model[1]	0,1,2,3	2,3	1,2,3,4	1,2	0,1,2,3,5	0	1
ICAM-2	Integrated Climate Assessment Model[2]	0,1,2,3	1,2	1,3,4	1,2	0,1,3	1,2,3	1,2
IIASA	International Institute for Applied Systems Analysis[3]	0	0	1	1	2	0	0
IMAGE 2.0*	Integrated Model to Assess the Greenhouse Effect[4]	0,1,2,3	3	0,2,3	2	1,2,3	1	1
MIT*	Massachusetts Institute of Technology[5]	0,1,2,3	2,3	1	2,3	0,2,3	1	0,1
PAGE	Policy Analysis of the Greenhouse Effect[6]	0,1	1,2	1	0	0,1,2,3,4	2	1
ProCAM*	Process Oriented Global Change Assessment Model[7]	0,1,2,3	2,3	1,2,3,4	2	0,2,3,5	1	1
TARGETS	Tool to Assess Regional and Global Environmental and Health Targets for Sustainability[8]	0,1,2,3,4	0	1,2,3,4	2	1,2,3,4	4	1,2
Policy Optimization Models								
AS/ExM	Adaptive Strategies Exploratory Model[9]	0	0	0	0	0	1	2
CETA	Carbon Emissions Trajectory Assessment[10]	0,1	0	1,2	0	0	0 or 1	0
Connecticut	Also known as the Yohe Model[11]	0	0	1	0	0	1	0
CRAPS	Climate Research and Policy Synthesis Model[12]	0	0	1	0	0	1	2
CSERGE	Centre for Social and Economic Research on the Global Environment[13]	0	0	1	0	0	1	0

Model		A	B	C	D	E	F	G
DIAM	Dynamics of Inertia and Adaptability Model[14]	0	0	1,2	0	0	0 or 1	0
DICE	Dynamic Integrated Climate and Economy Model[15]	0	0	1	0	0	0 or 1	0
FUND	The Climate Framework for Uncertainty, Negotiation and Distribution[16]	0,1	1	1,4	0	0,1,2,3,4	0 or 1	0
MARIA	Multiregional Approach for Resource and Industry Allocation[17]	0	0,1	1	0	0	0	0
MERGE 2.0	Model for Evaluating Regional and Global Effects of GHG Reductions Policies[18]	0,1	1	1,2	0	0	0 or 1	0
MiniCAM*	Mini Global Change Assessment Model[7]	0,1,2,3	2,3	1,2,3	2	0	0	1
PEF	Policy Evaluation Framework[19]	0,1	1,2	1	0	0	2	1
RICE	Regional DICE[20]	0	1	1	0	0	0	0
SLICE	Stochastic Learning Integrated Climate Economy Model[21]	0	1	1	0	0	1	2

* Models following the regionalized process-based approach described in this book

Key to codes:

A Forcings	B Geographical resolution	C Socio-economic dynamics	D Geophysical simulation**
0 CO_2	0 global	0 exogenous	0 global ΔT
1 other GHG	1 continental	1 economics	1 1-D ΔT and ΔP
2 aerosols	2 countries	2 technology choice	2 2-D ΔT and ΔP
3 land use	3 grids/basins	3 land use	3 2-D climate
4 other		4 demographic	

E Impact assessment	F Treatment of uncertainty	G Treatment of decision making
0 change in temperature	0 none	0 optimization
1 change in sea level	1 uncertainty	1 simulation
2 agriculture	2 variability	2 simulation with adaptive decisions
3 ecosystems	3 stochasticity	
4 health	4 cultural perspectives	
5 water		

** 1-D is 1-dimensional; ΔT is change in temperature; ΔP is change in precipitation

Source: Adapted from Weyant et al, 1996.[22]

SOME INTERNATIONAL DATA SOURCES OF INTEREST IN CLIMATE–IMPACT ASSESSMENT STUDIES

Table I *Data sources*

Type of data	Source	Spatial/temporal resolution	Content
Projections			
Population	IPCC[1]	7 regions and global/ totals in 2100	Total population (various projections)
Economic development	IPCC[1]	4 regions and global/ trends 1990–2100	GNP (average annual rate – various projections)
Gas and aerosol emissions	IPCC[1]	Global/annual rates 1990, 2025 and 2100	IS92a–f scenarios: CO_2 CH_4, N_2O, CFCs Halocarbons, SO_x
Radiative forcing	Wigley/Raper[2]	Global/annual up to 2100	IS92a–f scenarios and various asumptions
Climate change	NCAR[3]	Gridded (various resolutions)/daily, monthly and seasonal (time series or time slice up to 2100)	Equilibrium GCM (various models); Transient GCM (various models); Temperature, precipitation and other variables
Climate change	CRU[4]	Gridded (various resolutions) and globally averaged/monthly, seasonal and annual (time series or time slice up to 2100)	Equilibrium GCM (various models, including composite); Transient GCM (various models); 1-dimensional model (MAGICC); Temperature, precipitation and other variables
Sea-level rise	CRU[5]	Global/annual up to 2100	MAGICC (for any given emissions scenario)
Agriculture, forestry and fisheries	FAO[6]	Regional, global/ totals in 2010	Area, production, trade, consumption and other data

Current baseline

Population	UN[7]	National/annual	Total population/urban population (various projections)
Economic growth	World Bank[8]	National/annual	GNP, GDP
Climate	CDIAC[9]	Global stations/monthly (historical time series)	Temperature, precipitation, cloudiness, atmospheric pressure
Climate	UNEP/GRID[10]	Global 0.5° lat:lon grid/ 1931–60 period monthly means	Temperature, precipitation
Climate	CRU[11]	Global 5° lat:lon grid/ 1961–90 period monthly means	Temperature, precipitation
		Europe 0.5° lat:lon grid/ 1961–90 period monthly means	Temperature (max, min), precipitation, sunshine, windspeed, vapour pressure, rain days, frost days
Climate	ECMWF/ WCRP[12]	Global 2.5°, 1.125°, 0.5° lat:lon grid/ daily, monthly for individual years	Temperature, precipitation, atmospheric pressure
Land use/cover	UNEP/GRID[13]	Global 0.5° lat:lon grid/recent	Major ecosystem complexes based on maps and observations
Land use/cover	UNEP/GRID[14]	Global 1° lat:lon grid/ 1960–79	Predominant vegetation types, cultivation intensity and seasonal albedo based on maps
Land use/cover	UNEP/GRID[15]	Global 1° lat:lon grid	Wetlands (derived)
Agriculture, forestry and fisheries	FAO[6]	National, regional, global/ 1970, 1980, 1990	Area, production, trade, food supply and other data
General environment	UNEP[16]	National	Water, air, health and other environmental measures
Soil	UNEP/GRID[17]	Global 2 minute grid	FAO/UNESCO Soil Map of the World
Soil	UNEP/GRID[18]	Global 1° grid	Zobler soil type (based on UNESCO/FAO maps), soil texture, surface slope and other properties
Soil degradation	ISRIC[19]	Global	UNEP World Atlas of Desertification
Global Vegetation Index	UNEP/GRID[20]	78°N–55°S on 8.6 minute grid/1982–91	NOAA AVHHR Monthly Global Vegetation Index based on satellite data
Natural resources	WRI[21]	National/annual	Energy, raw materials, agriculture, forestry and many others
Human health	WHO[22]	National/annual	Distribution of and mortality from major diseases

Other data

Elevation/bathymetry	UNEP/GRID[23]	Global 5 minute grid	Integrated database derived from map information
Boundaries	UNEP/GRID[24]	Global (vector format)	World Databank II: Coastlines, islands, lakes, reefs, ice shelves, glaciers, rivers, canals, railways, administrative boundaries

Table II *Information about data sources*

Name (and media)	Source	Contents
ACCIS (Hardcopy)	UNEP[25]	Information services and computerized database
HEM (Hardcopy, Disk)	UNEP/GEMS[26]	Data banks; inventory of international research organizations and programmes; directory of environmental monitoring
INFOTERRA (Disk, Hardcopy)	UNEP[27]	Directory of information services
Master Directory (Network)	NASA[28]	Scientific data information service
CEOS-IDN (Network)	MECCA/NASA/ NASDA/ESA[29]	Directory of remotely sensed data

Notes for Table I:

1   Intergovernmental Panel on Climate Change (Leggett et al, 1992).
2   Wigley and Raper (1992).
3   National Center of Atmospheric Research, Boulder, Colorado, USA (Information from R Jenne and D Joseph).
4   Climatic Research Unit, University of East Anglia, Norwich, UK (Viner and Hulme, 1994; Hulme et al, 1995b).
5   As 4 (Wigley and Raper, 1992; Warrick et al, 1993).
6   Food and Agriculture Organization, United Nations (FAO, 1992; 1993).
7   United Nations (1991; 1992).
8   World Bank (1991).
9   Carbon Dioxide Information and Analysis Center, Oak Ridge, Tennessee, USA (Burtis, 1992).
10   United Nations Environment Programme/Global Resource Information Database (GRID-Geneva, 6 rue de la Gabelle, CH-1227 Carouge, Geneva, Switzerland). Climate data: Leemans and Cramer (1990).
11   As 4 (Jones, 1994; Hulme, 1994; Hulme et al, 1995a).
12   European Centre for Medium Range Weather Forecasting, Reading, UK/World Climate Research Programme (ECMWF, 1993).
13   As 10 (Olson et al, 1985).
14   As 10 (Matthews, 1983; 1985).
15   As 10 (Matthews and Fung, 1987).
16   United Nations Environment Programme (UNEP, 1987).
17   As 10 (FAO/UNESCO, various dates).

18   As 10 (Zobler, 1986).
19   International Soil Reference and Information Center (UNEP, 1992).
20   As 10 (Tarpley, 1991).
21   World Resources Institute (WRI, 1992).
22   World Health Organization (WHO, 1990).
23   As 10 (Haxby et al, 1983).
24   As 10 (CIA, 1972).

Notes for Table II:

25   ACCIS (1990).
26   Harmonization of Environmental Monitoring (UNEP/Global Environmental Monitoring System) (Fritz, 1990; Hicks, 1993).
27   International Referral System for Sources of Environmental Information (UNEP, 1987).
28   National Aeronautics and Space Administration, USA (Beier, 1991).
29   Commission on Earth Observing System-International Data Network (NASA/National Aeronautics and Space Development Agency/European Space Agency).

REFERENCES AND NOTES

CHAPTER 1

1   Carter, T R, Parry, M L, Harasawa, H, Nishioka, S (1994) *IPCC Technical Guidelines for Assessing Climate Change Impacts and Adaptations* Intergovernmental Panel on Climate Change, Department of Geography, University College London, UK and Centre for Global Environmental Research, Tsukuba, Japan, 59pp.
2   *Ibid.*

CHAPTER 2

1   Trenberth, K (1991) 'General characteristics of El Niño–Southern Oscillation' in Glantz, M, Katz, R and Nicholls, N (eds) *Teleconnections Linking Worldwide Climate Anomalies.* Cambridge University Press.
2   Berger, A (1980) The Milankovitch astronomical theory of palaeoclimatology: a modern review *Vistas in Astronomy* 24: 103–122.
3   Nicholls, N, Gruza, G V, Jouzel, J, Karl, T R, Ogallo, L A and Parker, D E (1996) 'Observed climate variability and change' in Houghton, J T, Meira Filho, L G, Callander, B A, Harris, N, Kattenberg, A and Maskell, K (eds) *Climate Change 1995: The Science of Climate Change* Cambridge University Press, pp 133–192.
4   Lamb, H H (1982) *Climate, History and the Modern World* Methuen, London and New York, 387 pp.
5   Wigley, T M L and Kelly, P M (1990) 'Holocene climate change, ^{14}C wiggles and variations in solar irradiance' *Phil Trans.* Royal Society, London A 330: pp 547–560.
6   For example, Huntley, B and Prentice, I C (1988) 'July temperatures in Europe from pollen data, 6000 years before present' *Science* 241: pp 687–690; Budyko, M and Izrael, Yu A (eds) (1987) *Anthropogenic Climatic Changes* L Gidrometeoizdat, 404 pp.
7   Lamb (1982) *Op cit*, Note 4.
8   Bradley, R S and Jones, P D (1993) 'Little Ice Age summer temperature variations: their nature and relevance to recent global warming trends' *The Holocene* 3: pp 367–376.
9   Nicholls et al (1996) *Op cit*, Note 3.
10 After Thompson, L G, Mosley-Thompson, E, Davis, M, Lin, P N, Yao, T, Dyergerov, M and Dai, J (1993) 'Recent warming: ice core evidence from tropical ice cores with emphasis on central Asia' *Global and Planetary Change* 7: pp 145–156.
11 Nicholls et al (1996) *Op cit*, Note 3.

12 Intergovernmental Panel on Climate Change (1996) 'Technical summary' in Houghton, J T, Meira Filho, L G, Callander, B A, Harris, N, Kattenberg, A and Maskell, K (eds) *Climate Change 1995: The Science of Climate Change* Cambridge University Press, pp 9–49.

13 Karl, T R, Jones, P D, Knight, R W, Kukla, G, Plummer, N, Razuvayev, V, Gallo, K P, Lindseay, J, Charlson, R J and Peterson, T C (1993) 'A new perspective on recent global warming: Asymmetric trends of daily maximum and minimum temperature' *Bulletin of the American Meteorological Society* 74: pp 1007–1023; Horton, E B (1995) 'Geographical distribution of changes in maximum and minimum temperatures' *Atmospheric Research* 37: pp 102–117.

14 Nicholls et al (1996) *Op cit*, Note 3.

15 Nicholson, S E (1994) 'Recent rainfall fluctuations in Africa and their relationship to past conditions', *Holocene* 4: pp 121–131.

16 Santer, B D, Wigley, T M L, Barnett, T P and Anyamba, E (1996) 'Detection of climate change and attribution of causes' in Houghton et al *Op cit*, Note 3, pp 407–443.

17 Trenberth, K E, Houghton, J T and Meira Filho, L G (1996) 'The climate system: an overview' in Houghton, J T, Meira Filho, L G, Callander, B A, Harris, N, Kattenberg A and Maskell, K (eds) *Climate Change 1995: The Science of Climate Change*. Cambridge University Press, pp 51–64. Note that this value of –19°C, which is commonly found in scientific publications, assumes the presence of clouds in the atmosphere and does not strictly apply to surface temperatures. In the absence of clouds, less radiation would be reflected to space and the surface temperature would be –6°C – see Houghton, J (1994) *Global Warming: The Complete Briefing* Lion Publishing, Oxford, 192 pp.

18 Intergovernmental Panel on Climate Change (1996) 'Technical summary' *Op cit*, Note 12.

19 Shine, K P, Fouquart, Y, Ramaswamy, V, Solomon, S and Srinivasan, J (1995) 'Radiative forcing' in Houghton, J T, Meira Filho, L G, Bruce, J, Hoesung Lee, Callander, B A, Haites, E, Harris, N and Maskell, K (eds) *Climate Change 1994. Radiative Forcing of Climate Change and An Evaluation of the IPCC IS92 Emission Scenarios*. Cambridge University Press, pp 163–203.

20 After Schimel, D and 26 others (1996) 'Radiative forcing of climate change' in Houghton et al (eds) *Climate Change 1995: The Science of Climate Change*. Cambridge University Press, pp 65–131.

21 Hansen, J, Lacis, A, Ruedy, R and Sato, M (1992) 'Potential climate impact of Mount Pinatubo eruption' *Geophysical Research Letters* 19: pp 215–218; Graf, H -F, Kirchner, I, Robock, A and Schult, I (1993) 'Pinatubo eruption winter climate effects: Model versus observations' *Climate Dynamics* 9: pp 81–93.

22 Smith, I N (1995) 'A GCM simulation of global climate interannual variability' *Journal of Climate* 8: pp 709–718.

23 Intergovernmental Panel on Climate Change (1996) 'Technical summary' *Op cit*, Note 12.

24 *Ibid*.

25 Leggett, J, Pepper, W J and Swart, R J (1992) 'Emissions scenarios for the IPCC: an update' in Houghton, J T, Callander, B A and Varney, S K (eds) *Climate Change 1992: The Supplementary Report to the IPCC Scientific Assessment*. Cambridge University Press, pp 69–95. These scenarios have since been slightly modified to account for the phase-out of CFCs – see Intergovernmental Panel on Climate Change (1996) *Op cit*, Note 12.

26 Intergovernmental Panel on Climate Change (1996) 'Technical summary' *Op cit*, Note 12.

27 *Ibid*.

28 *Ibid*.

29 Kattenberg, A, Giorgi, F, Grassl, H, Meehl, G A, Mitchell, J F B, Stouffer, R J, Tokioka, T, Weaver, A M and Wigley, T M L (1996) 'Climate models – projections of future climate' in Houghton et al (eds) *Op cit*, Note 3, pp 285–357.

30 Johns, T C, Carnell, R E, Crossley, J F, Gregory, J M, Mitchell, J F B, Senior, C A, Tett, S F B and Wood, R A (1997) 'The second Hadley Centre coupled ocean-atmosphere GCM: Model, description, spinup and validation', *Climate Dynamics* 13: pp 103–134.

31 Hadley Centre (1997) *Climate Change and its Impacts: A Global Perspective.* UK Meteorological Office, Bracknell, 16 pp.

32 Warrick, R A, LeProvost, C, Meier, M F, Oerleman, J, and Woodworth, P L (1996) 'Changes in sea level' in Houghton et al (1996) *Op cit*, Note 3, pp 359–405.

CHAPTER 3

1   Kates, R W (1985) 'The interaction of climate and society' in Kates, R W, Ausubel, J H and Berberian, M (eds) *Climate Impact Assessment.* SCOPE 27, John Wiley and Sons, Chichester, pp 3–36.

2   Climate Impact Assessment Program (CIAP) (1975) *Impacts of Climatic Change on the Biosphere*, Monograph 5. US Department of Transportation, Washington, DC.

3   Parry, M L and Carter, T R, (1988) 'The assessment of effects of climatic variations on agriculture' in Parry, M L, Carter, T R and Konijn, N T (eds) *The Impact of Climatic Variations on Agriculture, Volume 1: Assessments in Cool Temperate and Cold Regions; Volume 2: Assessments in Semi-Arid Regions.* Kluwer, Dordrecht, The Netherlands, pp 11–96.

4   Kates (1985) *Op cit*, Note 1.

5   National Defense University (NDU) (1980) *Crop Yields and Climatic Change to the Year 2000*, Vol 1. Fort Lesley J McNair, Washington, DC.

6   Williams, G D W, Fautley, R A, Jones, K H, Stewart, R B and Wheaton, E E (1988) 'Estimating effects of climatic change on agriculture in Saskatchewan, Canada' in Parry, M L, Carter, T R and Konijn, N T (eds) *The Impact of Climatic Variations on Agriculture, Volume 1, Assessments in Cool Temperate and Cold Regions.* Kluwer, Dordrecht, The Netherlands, pp 221–379.

7   Garcia, R (1981) *Drought and Man: The 1972 Case History. Vol 1: Nature Pleads Not Guilty.* Pergamon Press, New York.

8   Parry and Carter (1988) *Op cit*, Note 3.

9   Parry, M L, Carter, T R and Konijn, N T (eds) (1988) *The Impact of Climatic Variations on Agriculture, Volume 1: Assessments in Cool Temperate and Cold Regions; Volume 2: Assessments in Semi-Arid Regions.* Kluwer, Dordrecht, The Netherlands.

10 Williams et al (1988) *Op cit*, Note 6.

11 Parry and Carter (1988) *Op cit*, Note 3.

12 *Ibid.*

13 *Ibid.*

CHAPTER 4

1   Parry, M L and Read, N J (1988) *The Impact of Climatic Variability on UK Industry.* AIR Report No 1, Atmospheric Impacts Research Group, University of Birmingham, UK.

2   *Ibid.*

3   Smit, B (ed) (1993) *Adaptation to Climatic Variability and Change. Report of the Task Force on Climate Adaptation, The Canadian Climate Program.* Occasional Paper No 9, Department of Geography, University of Guelph, 53 pp.

4   *Ibid.*

5   Watson, R T, Zingwera, M C and Moss, R H (eds) (1996) *Climate Change 1995. Impacts, Adaptations and Mitigation of Climate Change: Scientific-Technical Analyses. Contribution of*

Working Group II to the Second Assessment Report of the IPCC. Cambridge University Press, 878 pp.

6   Parry, M L and Carter, T R (eds) (1984) Assessing *the Impact of Climatic Change in Cold Regions.* Summary Report SR-84–1, International Institute for Applied Systems Analysis, Laxenburg, Austria, 42 pp.

7   Kattenberg et al (1996) *Op cit,* Chapter 2, Note 29.

CHAPTER 5

1   Parry, M L and Carter, T R (eds) (1984) *Assessing the Impact of Climate Change in Cold Regions.* Summary Report SR-84–1, International Institute for Applied Systems Analysis, Laxenburg, Austria, 42 pp.

2   Parry and Carter (1988) *Op cit,* Chapter 3, Note 3.

3   Strzepek, K M and Smith, J B (eds) (1995) *As Climate Changes: International Impacts and Implications.* Cambridge University Press.

4   Parry and Carter (1984) *Op cit,* Note 1.

5   Parry et al (1988) *Op cit,* Chapter 3, Note 9.

6   Parry and Carter (1988) *Op cit,* Chapter 3, Note 3.

7   Parry, M L, Carter, T R and Hulme, M (1996) 'What is a dangerous climate change?' *Global Environmental Change* 6, pp 1–6.

8   Parry and Carter (1988) *Op cit,* Chapter 3, Note 3.

9   Strain, B R and Cure, J D (eds) (1985) *Direct Effects of Increasing Carbon Dioxide on Vegetation.* DOE/ER-0238, United States Department of Energy, Office of Energy Research, Washington, DC, 286 pp. Geign, S C, van de, Goudriaan, J and Berendse, F (eds) (1993) *Climate Change; Crops and Terrestrial Ecosystems.* Agrobiolosiche Thema's 9 CABO-DLO, Centre for Agrobiological Research, Wageningen, The Netherlands, 144 pp.

10 Parry and Carter (1988) *Op cit,* Chapter 3, Note 3.

11 Kattenberg et al (1996) *Op cit,* Chapter 2, Note 29.

12 WMO (1985) *Report of the WMO/UNEP/ICSU – SCOPE Expert Meeting on the Reliability of Crop–Climate Models for Assessing the Impacts of Climatic Change and Variability* WCP-90, World Meteorological Organization, Geneva, 31 pp.

13 WMO (1988) *Water Resources and Climatic Change: Sensitivity of Water Resource Systems to Climate Change and Variability.* WCAP-4, World Meteorological Organization, Geneva.

14 Bonan, G B (1993) 'Do biophysics and physiology matter in ecosystem models?', *Climatic Change,* 24: pp 281–285.

15 Warrick, R A (1984) 'The possible impacts on wheat production of a recurrence of the 1930s drought in the US Great Plains', *Climatic Change,* 6: pp 5–26.

16 *Ibid.*

17 UK Department of the Environment (1996) *Review of the Potential Effects of Climate Change in the United Kingdom.* HMSO, London.

18 *Ibid.*

19 Martens, W J M (1997) 'Health impacts of climate change and ozone depletion: an eco-epidemiological modelling approach' unpublished PhD thesis. Maastricht University, The Netherlands.

20 *Ibid.*

21 Smith, M (1992) *Expert Consultation on Revision of FAO Methodologies for Crop Water Requirements.* Land and Water Development Division, Food and Agriculture Organization, Rome, 60 pp.

22 Wolf, J, Evans, L G, Semenov, M A, Eckersten, H and Iglesias, A (1996) 'Comparison of wheat simulation models under climate change. I. Model calibration and sensitivity analysis', *Climate Research* 7: pp 253–270; Semenov, M A, Wolf, J, Evans, L G, Eckersten, H and Iglesias, A (1996) 'Comparison of wheat simulation models under climate change. II. Application of climate change scenarios', *Climate Research* 7: pp 271–281.

23 Kabat, P, Marshall, B, van den Broek, B J, Vos, J and van Keulen, H (eds) (1995) *Modelling and Parameterization of the Soil–Plant–Atmosphere System. A Comparison of Potato Growth Models*. Wageningen Pers, Wageningen, The Netherlands, 513 pp.

24 VEMAP Members (1995) 'Vegetation / ecosystem modeling and analysis project: Comparing biogeography and biogeochemistry models in a continental-scale study of terrestrial ecosystem responses to climate change and CO_2 doubling', *Global Biogeochemical Cycles* 9: pp 407–437.

25 IBSNAT (1989) *Decision Support System for Agrotechnology Transfer Version 2 1 (DSSAT V2.1)*. International Benchmark Sites Network for Agrotechnology Transfer, Department of Agronomy and Soil Science, College of Tropical Agriculture and Human Resources, University of Hawaii, Honolulu.

26 Parry, M L, Hossell, J E, Jones, P J, Rehman, T, Tranter, R B, Marsh, J S, Rosenzweig, C, Fischer, G, Carson, I G, and Bunce, R G H (1996) 'Integrating global and regional analyses of the effects of climate change: a case study of land-use in England and Wales', *Climatic Change* 32: pp 185–198.

27 Adams, R M, Rosenzweig, C, Peart, R M, Ritchie, J T, McCarl, B A, Glyer, J D, Curry, R B, Jones, J W, Boote, K J and Allen, L H, Jr (1990) 'Global climate change and US agriculture', *Nature*, 345(6272). pp 219-222.

28 Lovell, C A K and Smith, V K (1985) 'Microeconomic analysis' in Kates, R W, Ausubel, J H and Berberian, M (eds) *Climate Impact Assessment: Studies of the Interaction of Climate and Society* SCOPE 27, John Wiley and Sons, Chichester, pp 293-321.

29 Parry et al (1996) *Op cit*, Note 26.

30 Williams et al (1988) *Op cit*, Chapter 3, Note 6.

31 *Ibid*.

32 Adams et al (1990) *Op cit*, Note 27.

33 Scheraga, J D, Leary, N, Goettle, R, Jorgenson, D and Wilcoxen, P (1993) 'Macroeconomic modeling and the assessment of climate change impacts' in Kaya, Y, Nakicenovic, N, Nordhaus, W D and Toth, F L (eds) *Costs, Impacts and Benefits of CO_2 Mitigation* Collaborative Paper Series CP93-2, International Institute for Applied Systems Analysis, Laxenburg, Austria.

34 Rosenzweig, C R, Parry, M L, Fischer, G and Frohberg, K (1993) *Climate Change and World Food Supply* Research Report No 3. Environmental Change Unit, University of Oxford, 28 pp.

35 Cline, W (1992) *The Economics of Global Warming*. Institute for International Economics, Washington, D C; Fankhauser, S (1994) 'The social costs of greenhouse gas emissions: an expected value approach' *Energy Journal* 15(2): pp 157–184.

36 Rosenzweig et al (1993) *Op cit*, Note 34.

37 Nordhaus, W D (1994) *Managing the Global Commons: The Economics of Global Change*. MIT Press, Cambridge, US.

38 Peck, S C and Teisberg, T J (1992) 'CETA: A model for carbon emissions trajectory assessment', *The Energy Journal* 13: pp 55–77; Peck, S C and Teisberg, T J (1994) 'Optimal carbon emissions trajectories when damages depend on the rate or level of warming', *Climatic Change* 30: pp 289–314.

39 Manne, A S, Mendelsohn, R and Richels, R G (1993) 'MERGE: A model for evaluating regional and global effects of GHG reduction policies', *Energy Policy* 23: pp 17–34.

40 Weyant, J, Davidson, O, Dowlatabadi, H, Edmonds, J, Grubb, M, Parson, E A, Richels, R, Rotmans, J, Shukla, P R, Tol, R S J, Cline, W and Fankhauser, S (1996) 'Integrated

assessment of climate change: an overview and comparison of approaches and results' in Bruce, J P, Lee, H and Haites, E F (eds) *Climate Change 1995. Economic and Social Dimensions of Climate Change. Contribution of Working Group III to the Second Assessment Report of the Intergovernmental Panel on Climate Change.* Cambridge University Press, pp 367–396.

41 Rotmans, J (1990) *IMAGE: an Integrated Model to Assess the Greenhouse Effect.* Kluwer, Dordrecht; Alcamo, J (ed) (1994) *IMAGE 2.0: Integrated Modelling of Global Climate Change.* Kluwer, Dordrecht, The Netherlands, 321 pp (reprinted from *Water, Air and Soil Pollution* 76).

42 Rotmans, J, Hulme, M and Downing, T E (1994) 'Climate change implications for Europe. An application of the ESCAPE model', *Global Environmental Change* 4(2): pp 97–124; Hulme, M, Raper, S and Wigley, T M L (1995) 'An integrated framework to address climate change (ESCAPE) and further developments of the global and regional climate modules (MAGICC)', *Energy Policy* 23(4/5): pp 347–355.

43 Alcamo, J (ed) (1994) *IMAGE 2.0: Integrated Modeling of Global Climate Change.* Kluwer, Dordrecht, The Netherlands, 321 pp (reprinted from *Water, Air and Soil Pollution* 76); Alcamo, J, Kreileman, E and Leemans, R (eds) (1996) 'Integrated scenarios of global change: results from the IMAGE 2 model', *Global Environmental Change* 6: pp 255–394.

44 Weyant et al (1996) *Op cit*, Note 40.

45 Emery, K O and Aubrey, D G (1991) *Sea Levels, Land Levels and Tide Gauges.* Springer Verlag, New York, 237 pp.

46 Alcamo (1994); Alcamo et al (1996) *Op cit*, Note 43.

47 Manabe, S and Wetherald, R T (1987) 'Large scale changes of soil wetness induced by an increase in atmospheric carbon dioxide', *Journal of Atmospheric Science* 44: pp 1211–1235.

48 Vloedbeld, M and Leemans, R (1993) 'Quantifying feedback processes in the response of the terrestrial carbon cycle to global change: The modelling approach of IMAGE-2', *Water, Air and Soil Pollution* 70: pp 615–628.

49 Bergthorsson, P, Björnsson, H, Dyrmundsson, O R, Gudmundsson, B, Helgadottir, A and Jonmundsson, J V (1988) 'The effects of climatic variations on agriculture in Iceland' in Parry, M L, Carter, T R, and Konijn, N T (eds) *The Impacts of Climatic Variations on Agriculture. Volume 1, Effects in Cool Temperate and Cold Regions.* Kluwer, Dordrecht, The Netherlands, pp 381–509.

50 National Defense University (NDU) 1978 Climate Change to the Year 2000. Washington, DC, Fort Lesley J McNair; National Defense University (NDU) (1980) *Crop Yields and Climate Change to the Year 2000.* Vol 1. Fort Lesley J McNair Washington, DC.

51 Stewart, T R and Glantz, M H (1985) 'Expert judgment and climate forecasting: A methodological critique of Climate Change to the Year 2000', *Climatic Change* 7: pp 159–183.

52 Parry and Carter (1988) *Op cit*, Chapter 3, Note 3.

53 Hansen, J, Russell, G, Rind, D, Stone, P, Lacis, A, Lebedeff, S, Ruedy, R and Travis, L (1983) 'Efficient three-dimensional global models for climate studies: Models I and II', *Monthly Weather Review* 111: pp 609–662.

54 Kuikka, S and Varis, O (1997) 'Uncertainties of climate change impacts in Finnish watersheds: a Bayesian network analysis of expert knowledge', *Boreal Environment Research* 2: pp 109–128; Varis, O and Kuikka, S (1997) 'BENE-EIA: A Bayesian approach to expert judgment elicitation with case studies on climate change impacts on surface waters', *Climatic Change* 37: pp 539–563.

55 Yen, Y and Cohen, S J (1994) 'Identifying regional goals and policy concerns associated with global climate change', *Global Environmental Change* 4(3): pp 246–260; Cohen S J (ed) (1997) *Mackenzie Basin Impact Study (MBIS).* Final report. Atmospheric Environment Service, Environment Canada, 372 pp.

CHAPTER 6

1   Scott, M J (1993) 'Impacts of Climate Change on Human Settlements: Guidelines for Assessing Impacts'. Draft Mimeo submitted to the UNEP-Canada Workshop on Impacts and Adaptation to Climate Variability and Change, Toronto, Canada, 29 November–3 December 1993, 3 pp.

2   *Ibid.*

3   Downing, T E (1992) *Climate change and vulnerable places: global food security and country studies in Zimbabwe, Kenya, Senegal and Chile.* Research Report No 1, Environmental Change Unit, University of Oxford, 54 pp.

4   *Ibid.*

5   Pittock, A B (1993) 'Climate scenario development' in Jakeman, A J, Beck, M B and McAleer, M J (eds) *Modelling Change in Environmental Systems.* John Wiley and Sons, pp 481–503.

6   Whetton, P H, Hennessy, K J, Pittock, A B, Fowler, A M and Mitchell, C D (1992) 'Regional impact of the enhanced greenhouse effect on Victoria' in CSIRO Division of Atmospheric Research *Annual Report 1991–92.* Commonwealth Scientific and Industrial Research Organisation, Mordialloc, 64 pp.

7   For example, see Parry, M L (1978) *Climatic Change, Agriculture and Settlement.* Dawson, Folkestone, 214 pp.

8   Major works of this kind include Lamb, H H (1977) *Climate: Present, Past and Future, Volume 2: Climatic History and the Future.* Methuen, London, 835 pp; Pfister, C (1984) *Das Klima der Schweiz und Seine Bedeutung in der Geschichte von Befolkenung und Landwirtschaft.* 2 volumes. Haupt, Bern.

9   Manley, G (1974) 'Central England Temperatures: monthly means 1659 to 1973', *Quarterly Journal of the Royal Meteorological Society* 100: pp 389–405.

10 The series is updated by the UK Hadley Centre; see Parker, D E, Legg, T P and Folland, C K (1992) 'A new daily Central England Temperature Series: 1772–1991', *International Journal of Climatology* 12: pp 317–342.

11 *Ibid.*

12 Farhar-Pilgrim, B (1985) 'Social analysis' in Kates, R W, Ausubel, J H and Berberian M (eds) *Climate Impact Assessment: Studies of the Interaction of Climate and Society.* SCOPE 27, John Wiley and Sons, Chichester, pp 323–350.

13 Whyte, A V T (1985) 'Perception' in Kates, R W, Ausubel, J H and Berberian, M (eds) *Climate Impact Assessment: Studies of the Interaction of Climate and Society.* SCOPE 27, John Wiley and Sons, Chichester, pp 403–436.

14 Akong'a, J, Downing, T E, Konijn, N T, Mungai, D N, Muturi, H R and Potter, H L (1988) 'The effects of climatic variations on agriculture in central and eastern Kenya' in Parry, M L, Carter, T R, and Konijn, N T (eds) *The Impact of Climatic Variations on Agriculture, Volume 2: Assessments in Semi-Arid Regions.* Kluwer, Dordrecht, The Netherlands, pp 121–270; Gadgil, S, Huda, A K S, Jodha, N S, Singh, R P and Virmani, S M (1988) 'The effects of climatic variations on agriculture in dry tropical regions of India' in Parry, M L, Carter, T R and Konijn, N T (eds) *The Impact of Climatic Variations on Agriculture, Volume 2: Assessments in Semi-Arid Regions.* Kluwer, Dordrecht, The Netherlands, pp 495–578.

15 Jones, P and Hulme, M (1997) 'The changing temperature of "central England"' in Hulme, M and Barrow, E (eds) *Climates of the British Isles.* Routledge, London, pp 173–196.

16 Manley (1974) *Op cit*, Note 9.

17 See Note 10.

18 Hulme, M (1994) 'The cost of climate data – a European experience', *Weather* 49: pp 168–175.

19 Carter, T R, Konijn, N T and Watts, R G (1988) 'The choice of first-order impact models' in Parry, M L, Carter, T R and Konijn, N T (eds) *The Impact of Climatic Variations on Agriculture. Volume 1. Effects in Cool, Temperate and Cold Regions*. Kluwer, Dordrecht, The Netherlands, pp 97–123.

CHAPTER 7

1   Parry, M L and Read, N J (1988) (eds) *The Impact of Climatic Variability on UK Industry*. AIR Report 1. Atmospheric Impacts Research Group, University of Birmingham, 71 pp.
2   Carter, T R, Parry, M L, Harasawa, H and Nishioka, S (1994) *IPCC Technical Guidelines for Assessing Climate Change Impacts and Adaptations*. Intergovernmental Panel on Climate Change, Department of Geography, University College London, UK and Center for Global Environmental Research, National Institute for Environmental Studies, Tsukuba, Japan, 59 pp.
3   Leggett, J, Pepper, W J and Swart, R J (1992) 'Emissions scenarios for the IPCC: an update' in Houghton, J T, Callander, B A and Varney, S K (eds) *Climate Change 1992: The Supplementary Report to the IPCC Scientific Assessment*. Cambridge University Press, pp 69–95.
4   Wigley, T M L and Raper, S C B (1992) 'Implications for climate and sea level of revised IPCC emissions scenarios', *Nature*, 357: pp 293–300.
5   Leggett et al (1992) *Op cit*, Note 3.
6   Wigley and Raper (1992) *Op cit*, Note 4.
7   Carter et al (1994) *Op cit*, Note 2.
8   Leggett et al (1992) *Op cit*, Note 3.
9   Wigley and Raper (1992) *Op cit*, Note 4.
10 Budyko, M I (1989) 'Empirical estimates of imminent climatic changes', *Soviet Meteorology and Hydrology* 10: pp 1–8.
11 For example, see Lough, J M, Wigley, T M L and Palutikof, J P (1983) 'Climate and climate impact scenarios for Europe in a warmer world', *Journal of Climate and Applied Meteorology* 22: pp 1673–1684.
12 Warrick, R A (1984) 'The possible impacts on wheat production of a recurrence of the 1930s drought in the US Great Plains', *Climatic Change*, 6: pp 5–26; Williams, G D V, Fautley, R A, Jones, K H, Stewart, R B, and Wheaton, E E (1988) 'Estimating Effects of Climatic Change on Agriculture in Saskatchewan, Canada' in Parry, M L, Carter, T R and Konijn, N T (eds) *The Impact of Climatic Variations on Agriculture, Volume 1: Assessments in Cool, Temperate and Cold Regions*. Kluwer, Dordrecht, The Netherlands, pp 219–379; Rosenberg, N J (ed) (1993) 'Towards an integrated impact assessment of climate change: the MINK study', *Climatic Change* (Special Issue) 24: pp 1–173
13 Parry, M L and Carter, T R (1988) 'The assessment of effects of climatic variations on agriculture: a summary of results from studies in semi-arid regions' in Parry, M L, Carter, T R and Konijn, N T (eds) *The Impact of Climatic Variations on Agriculture, Volume 2: Assessments in Semi-Arid Regions*. Kluwer, Dordrecht, The Netherlands, pp 9–60.
14 Rosenberg, N J, Crosson, P R, Frederick, K D, Easterling, W E, III, McKenney, M S, Bowes, M D, Sedjo, R A, Darmstadter, J, Katz, L A and Lemon, K M (1993) 'Paper 1. The MINK methodology: background and baseline', *Climatic Change* 24: pp 7–22.
15 Mitchell, J F B, Manabe, S, Meleshko, V and Tokioka, T (1990) 'Equilibrium climate change – and its implications for the future' in Houghton, J T, Jenkins, G J and Ephraums, J J (eds) *Climate Change: The IPCC Scientific Assessment*. Report of Working Group I of the Intergovernmental Panel on Climate Change, Cambridge University Press, pp 131–164; Gates, W L, Mitchell, J F B, Boer, G J, Cubasch, U and Meleshko, V P (1992) 'Climate

modelling, climate prediction and model validation' in Houghton, J T, Callander, B A and Varney, S K (eds) *Climate Change 1992: The Suplementary Report to the IPCC Scientific Assessment*. Cambridge University Press, pp 97–134.

16 Kattenberg, A, Giorgi, F, Grassl, H, Meehl, G A, Mitchell, J F B, Stouffer, R J, Tokioka, T, Weaver, A J and Wigley, T M L (1996) 'Climate models – projections of future climate' in Houghton, J T, Meira Filho, L G, Callander, B A, Harris, N, Kattenberg, A and Maskell, K (eds) *Climate Change 1995: The Science of Climate Change. Contribution of Working Group I to the Second Assessment Report of the Intergovernmental Panel on Climate Change*. Cambridge University Press, pp 285–357.

17 *Ibid.*

18 Two examples of scenario software are Hulme, M, Jiang, T and Wigley, T (1995) *SCENGEN: A Climate Change SCENario GENerator. Software User Manual, Version 1.0.* Climatic Research Unit, University of East Anglia, Norwich, 38 pp and Jones, R N (1996) *OzClim – A Climate Scenario Generator and Impacts Package for Australia*. CSIRO Division of Atmospheric Research, Commonwealth Scientific and Industrial Research Organisation, Aspendale, Australia.

19 For example, Bultot, F, Coppens, A, Dupriez, G L, Gellem, D and Meulenberghs, F (1988) 'Repercussions of a CO_2 doubling on the water cycle and on the water balance – a case study for Belgium', *Journal of Hydrology* 99: pp 319–347; Croley, J E,II (1990) 'Laurentian Great Lakes double-CO_2 climate change hydrological impacts', *Climatic Change* 17: pp 27–47.

20 Parry, M L and Carter, T R (1988) 'The assessment of effects of climatic variations on agriculture. aims, methods and summary of results' in Parry, M L, Carter, T R and Konijn, N T (eds) *The Impact of Climatic Variations on Agriculture. Volume 1: Assessments in Cool, Temperate and Cold Regions*. Kluwer, Dordrecht, The Netherlands, pp 11–95.

21 Kattenberg et al (1996) *Op cit*, Note 16.

22 *Ibid.*

23 Hulme, M and Brown, O (1998) 'Portraying climate scenario uncertainties in relation to tolerable regional climate change'. *Climate Research* (in press).

24 Mitchell, J F B, Johns, T C, Gregory, J M and Tett, S F B (1995) 'Climate response to increasing levels of greenhouse gases and sulphate aerosols', *Nature* 376: pp 501–504; Johns, T C, Carnell, R E, Crossley, J F, Gregory, J M, Mitchell, J F B, Senior, C A, Tett, S F B and Wood, R A (1997) 'The second Hadley Centre coupled ocean-atmosphere GCM: model description, spinup and validation', *Climate Dynamics* 13: pp 103–134.

25 Jones, P D (1994) 'Hemispheric surface air temperature variability – a reanalysis and an update to 1993', *Journal of Climate* 7: pp 1794–1802

26 Hulme, M (1994) 'Validation of large-scale precipitation fields in general circulation models' in Desbois, M and Désalmand, F (eds) *Global Precipitation and Climate Change*. Springer-Verlag, Berlin, pp 387–405

27 Mitchell, J F B, Johns, T C and Davis, R A (1998) 'Towards the construction of climate change scenarios', *Climatic Change* (in press).

28 Hulme and Brown (1998) *Op cit*, Note 23.

29 *Ibid.*

30 Mearns, L O, Rosenzweig, C and Goldberg, R (1996) 'The effect of changes in daily and interannual climatic variability on CERES-Wheat: A sensitivity study', *Climatic Change* 32: pp 257–292; Mearns, L O, Rosenzweig, C and Goldberg, R (1997) 'Mean and variance changes in climate scenarios: methods, agricultural applications, and measures of uncertainty', *Climatic Change* 35: pp 367–396; Semenov, M A and Porter, J R (1995) 'Climatic variability and modelling of crop yields', *Agricultural and Forest Meteorology* 73: pp 265–283.

31 See also Wilks, D S (1992) 'Adapting stochastic weather generation algorithms for climate change studies', *Climatic Change* 22: pp 67–84; Semenov, M A and Barrow, E M (1997) 'Use of a stochastic weather generator in the development of climate change scenarios', *Climatic Change* 35: pp 397–414.

CHAPTER 8

1 Muchena, P and Iglesias, A (1995) 'Vulnerability of maize yields to climate change in different farming sectors in Zimbabwe' in *Climate Change and Agriculture: Analysis of Potential International Impacts*. ASA Special Publication No 59: pp 229–240.

2 Kalkstein, L S and Tan, G, (1995) 'Human Health' in Strzepek, K M and Smith, J B (eds) *As Climate Changes: International Impacts and Implications*. Cambridge University Press, pp 124–145.

3 Holten, J I and Carey, P D (1992) *Responses of Climate Change on Natural Terrestrial Ecosystems in Norway*. NINA Forskningsraport 29, Norwegian Institute for Nature Research, Trondheim, Norway.

4 Muchena and Iglesias (1995) *Op cit*, Note 1.

5 Kalkstein and Tan (1995) *Op cit*, Note 2.

6 Holten and Carey (1992) *Op cit*, Note 3.

7 Kirschbaum, M U F, Fischlin, A, Cannell, M G R, Cruz, R V O, Galinski, W and Cramer, W P (1996) 'Climate change impacts on forests' in Watson, R T, Zinyowera, M C and Moss, R H (eds) *Climate Change 1995. Impacts, Adaptations and Mitigation of Climate Change: Scientific-Technical Analyses. Contribution of Working Group II to the Second Assessment Report of the Intergovernmental Panel on Climate Change*, Cambridge University Press, pp 95–129.

8 Prentice, I C, Cramer, W P, Harrison, S P, Leemans, R, Monserud, R A and Solomon, A M (1992) 'A global biome model based on plant physiology and dominance, soil properties and climate', *Journal of Biogeography* 19: pp 117–134; Prentice, I C, Sykes, M T, Lautenschlager, M, Harrison, S P, Denissenko, O and Bartlein, P J (1994) 'Modelling global vegetation patterns and terrestrial carbon storage at the last glacial maximum', *Global Ecology and Biogeography Letters* 3: pp 67–76.

9 Wetherald, R T and Manabe, S (1986) 'An investigation of cloud cover change in response to thermal forcing', *Climatic Change* 8: pp 5–23.

10 Rosenzweig, C, Parry, M L, and Fisher, G (1995) 'World Food Supply' in Strzepek, K M and Smith, J B (eds) *As Climate Changes: International Impacts and Implications*, Cambridge University Press.

11 Prentice et al (1992) *Op cit*, Note 8.

12 Rosenzweig et al (1995) *Op cit*, Note 10.

13 Rosenzweig, C and Parry, M L (1994) 'Potential impacts of climate change on world food supply', *Nature* 367: pp 133–8.

14 Parry, M L, (1976) 'The significance of the variability of summer warmth in upland Britain', *Weather* 31: pp 212–217.

15 UK Department of the Environment (1996) *Review of the Potential Effects of Climate in the United Kingdom*. HMSO, London.

16 Rosenzweig and Parry (1994) *Op cit*, Note 13.

17 UK Department of the Environment (1996) *Op cit*, Note 15.

18 Parry, M L, and Carter, T R (1985) 'The effect of climatic variations on agricultural risk', *Climatic Change* 7: pp 95–110.

19 Stakhiv, E Z, Ratick, S J, and Du, W (1991) 'Risk-cost aspects of sea level rise and climate change in the evaluation of shore protection projects' in Ganoulis, J (ed) *Water Resources*

Engineering Risk Assessment. NATO ASI Series, Springer-Verlag, Berlin, pp 311–335; Stakhiv, E Z, (1993) 'Water resources planning and management under climate uncertainty' in Ballentine, T M and Stakhiv, E Z (eds) *Proceedings of the First National Conference on Climate Change and Water Resources Management.* IWR Report 93-R–17 Institute of Water Resources, US Army Corps of Engineers, Washington DC, IV-20-IV-35.

20   Parry and Carter (1985) *Op cit*, Note 18.

21   *Ibid.*

22   Pearce, D W (1993) *Economic Values and the Natural World.* Earthscan, London, 144 pp.

23   Holling, C S (ed) (1978) *Adaptive Environmental Assessment and Management.* John Wiley and Sons, Chichester, 377 pp.

24   Martin, P and Lefebvre, M (1995) '9 to 5. 9 approaches to tackle 5 aspects of climate change', *Climatic Change* 25: pp 421–438.

25   *Ibid.*

26   Saarikko, R A and Carter, T R (1996) 'Estimating the development and regional thermal suitability of spring wheat in Finland under climatic warming', *Climate Research* 7: pp 243–252.

27   *Ibid.*

CHAPTER 9

1   For more information on this topic, the reader is directed to parallel work by Working Group III of the Intergovernmental Panel on Climate Change in Bruce, J P, Lee, H and Haites, E F (eds) (1996) *Climate Change 1995. Economic and Social Dimensions of Climate Change.* Contribution of Working Group III to the Second Assessment Report of the Intergovernmental Panel on Climate Change. Cambridge University Press, 448 pp.

2   Beuker, E (1994) 'Long-term effects of temperature on the wood production of *Pinus sylvestris* L and *Picea abies* L Karst in old provenance experiments', *Scandinavian Journal of Forest Research* 9: pp 34–45.

3   Rosenzweig, C R and Parry, M L, (1994) 'Potential impact of climate change on world food supply', *Nature* 367: pp 133–8.

4   *Ibid.*

5   Strzepek, K M, Onyeji, S C, Saleh, M and Yates, D (1995) 'An assessment of integrated climate change impact on Egypt' in Strzepek, K M and Smith, J (eds) *As Climate Changes: International Impacts and Implications.* Cambridge University Press, pp 180–200.

6   Carter, T R, Parry, M L, Harasawa, H, and Nishioka, S (1994) *IPCC Technical Guidelines for Assessing Climate Change Impacts and Adaptations.* Department of Geography, University College London and Centre for Global Environmental Research, Tsukuba, Japan, 59 pp.

7   Rosenzweig and Parry (1994) *Op cit*, Note 3.

8   *Ibid.*

9   Strzepek, K M and Smith, J B (1995) *As Climate Changes: International Impacts and Implications.* Cambridge University Press, 231 pp.

10   Strzepek et al (1995) *Op cit*, Note 5.

11   Wilson, C A and Mitchell, J F B (1987) 'A doubled CO_2 climate sensitivity experiment with a global climate model including a simple ocean', *Journal of Geophysical Research* 92(13): pp 315–343.

12   Hansen, J, Russell, G, Rind, D, Stone, P, Lacis, A, Lebedeff, S, Ruedy, R and Travis, L (1983) 'Efficient three-dimensional global models for climate studies: Models I and II', *Monthly Weather Review* 111: pp 609–662.

13   Manabe, S and Wetherald, R T (1987) 'Large scale changes of soil wetness induced by an increase in atmospheric carbon dioxide', *Journal of Atmospheric Science* 44: pp 1211–1235.

14 Carter et al (1994) *Op cit*, Note 6.

15 Stakhiv, E Z (1994) 'Water resources planning of evaluation principles applied to ICZM' in *Proc. Preparatory Workshop on Integrated Coastal Zone Management and Responses to Climate Change*. World Coast Conference 1993, New Orleans, Louisiana.

16 *Ibid.*

17 Intergovernmental Panel on Climate Change (1994) *Preparing to Meet the Coastal Challenges of the 21st Century*. Conference Report. World Coast Conference 1993. Nordwijk, The Netherlands.

18 Burton, I, Kates, R W and White, G F (1993) *Environment as Hazard*. Guildford Press, New York.

19 IPCC (1994) *Op cit*, Note 17.

20 Hotthus, P, Crawford, M, Makroro, C and Sullivan, S (1992) 'Vulnerability assessment for accelerated sea level rise case study: Majuro Atoll, Republic of the Marshall Islands', *SPREP Reports and Studies Series* No 60, Apia, Western Samoa, 107 pp.

21 From Burton, I (1996) 'Assessment of adaptation to climate change' in *Handbook on Methods for Climate Change Impacts Assessment and Adaptation Strategies*. Draft Version 1.3. United Nations Environmental Programme and Institute for Environmental Studies, Amsterdam, The Netherlands.

22 Johda, N S and Mascarenhas, A C (1985) 'Adjustment in self-provisioning societies' in Kates, R W, Ausubel, J H and Berberian, M (eds) *Climate Impact Assessment: Studies of the Interaction of Climate and Society*. SCOPE 27, John Wiley and Sons, Chichester pp 437–464.

23 Akong'a, J, Downing, T E, Konijn, N T, Mungai, D N, Muturi, H R, and Potter, H L (1988) 'The effects of climatic variations on agriculture in central and eastern Kenya' in Parry, M L, Carter, T R, and Konijn, N T (eds) *The Impacts of Climatic Variations on Agriculture, Volume 2: Assessments in Semi Arid Regions*. Kluwer, Dordrecht, The Netherlands pp 121–270.

24 Stakhiv, E Z (1993) 'Water resources planning and management under climate uncertainty' in Ballentime, T M and Stakhiv, E Z (eds) *Proceedings of the First National Conference on Climate Change and Water Resources Management*. IWR Report 93–R–17, Institute of Water Resources, US Army Corps of Engineers, Washington DC IV-20-IV-35.

25 Akong'a et al (1988) *Op cit*, Note 23.

26 Burton (1996) *Op cit*, Note 21.

27 Rosenberg, N J (ed) (1993) 'Towards an integrated impact assessment of climate change: the MINK study', *Climatic Change* 24: pp 1–173.

28 Carter, T R (1996) 'Assessing climate change adaptations: the IPCC Guidelines' in Smith, J B, Bhatti, N, Menzhulin, G V, Benioff, R, Campos, M, Jarrow, B, Rijsberman, F, Budyko, M I and Dixon, R K (eds) *Adapting to Climate Change: An International Perspective*. Springer-Verlag, New York, pp 27–43.

29 Carter et al (1994) *Op cit*, Note 6.

30 Goicoechea, A, Hansen, D and Duckstein, L (1982) *Multiobjective Decision Analysis with Engineering and Business Application*. John Wiley and Sons, New York, 520 pp; and see, for example, Chankong, V, and Haimes, Y, (1983) *Multiobjective Decision Making: Theory and Methodology*. Elsevier, New York, 406 pp.

31 Stakhiv (1994) *Op cit*, Note 15.

32 Rosenberg (1993) *Op cit*, Note 27.

33 Toth, F L (1989) 'Policy exercises', *IIASA Research Report RR-89-2*. Reprinted from *Simulation and Games* 19 (3). International Institute for Applied Systems Analysis, Laxenburg, Austria, 43 pp.

34 Parry, M L, Blantran de Rozari, M, Chong, A L, and Panich, S, (1992) *The Potential Socio-economic Effects of Climate Change in South-East Asia*. United Nations Environment Programme, Nairobi, 126 pp.

CHAPTER 10

1   Carter et al (1994) *Op cit*, Chapter 1, Note 1.
2   For example, see Gilpin, A (1995) *Environmental Impact Assessment: Cutting Edge for the 21st Century*. Cambridge University Press, 198 pp.

APPENDIX 1

1   Morita, T, Kaihuma, M, Harasawa, H, Kai, Dong-Kumand, L and Matsuoka, Y (1994) *Asian-Pacific Integrated Model for Evaluating Policy Options to Reduce GHG Emissions and Global Warming Impacts, Interim Report*. National Institute for Environmental Studies, Tsukuba, Japan.

2   Dowlatabadi, H (1995) *Integrated Assessment Climate Assessment Model 2.0, Technical Documentation*. Department of Engineering and Public Policy, Carnegie Mellon University, Pittsburgh.

3   WEC (World Energy Council) and IIASA (International Institute for Applied Systems Analysis) (1995) *Global Energy Perspectives to 2050 and Beyond*. World Energy Council, London.

4   Alcamo, J (ed) (1994) *IMAGE 2.0: Integrated Modeling of Global Climate Change*, Kluwer, Dordrecht, The Netherlands, 321 pp (reprinted from *Water, Air and Soil Pollution* 76).

5   MIT (Massachusetts Institute of Technology (1994) *Joint Program on the Science and Technology of Global Climate Change*. Center for Global Change Science and Center for Energy and Environmental Policy Research, Cambridge, MA.

6   Commission of the European Communities (1992) *PAGE User Manual*. Brussels.

7   Edmonds, J, Pitcher, H, Rosenberg, N and Wigley, T (1994) *Design for the Global Change Assessment Model*. Proceedings of the International Workshop on Integrative Assessment of Mitigation, Impacts and Adaptation to Climate Change, International Institute for Applied Systems Analysis, Laxenburg, Austria, 13–15 October.

8   Rotmans, J, van Asselt, M B A, de Bruin, A J, den Elzen, M G J, de Greef, J, Hilderink, H, Hoekstra, A Y, Janssen, M A, Koster, H W, Martens, W J M, Niessen, L W and de Vries, H J M (1994) *Global Change and Sustainable Development: A Modelling Perspective for the Next Decade*. National Institute of Public Health and Environmental Protection, Bilthoven, The Netherlands.

9   Lempert, R J, Schlesinger, M E and Hammitt, J K (1994) 'The impact of potential abrupt climate changes on near-term policy choices', *Climatic Change* 26: pp 351–376.

10 Peck, S C and Teisberg, T J (1992) 'CETA: A model for carbon emissions trajectory assessment', *The Energy Journal* 13: pp 55–77.

11 Yohe, G (1995) *Exercises in Hedging against Extreme Consequences of Global Change and the Expected Value of Information*. Department of Economics, Wesleyan University, Wesleyan, CT.

12 Jain, A, Kheshgi, H and Wuebbles, D (1994) *Integrated Science Model for Assessment of Climate Change*. Lawrence Livermore Laboratory, Livermore, CA.

13 Maddison, D (1995) 'A cost-benefit analysis of slowing climate change' *Energy Policy* 23: pp 337–346.

14 Grubb, M, Duong, M H and Chapuis, T (1995) 'The economics of changing course', *Energy Policy* 23: pp 417–432.

15 Nordhaus, W D (1994) *Managing the Global Commons: The Economics of Climate Change*. MIT Press, Cambridge, MA.

16 Tol, R S J, van der Burg, T, Jansen, H M A and Verbruggen, H (1995) 'The climate fund – some notions on the socio-economic impacts of greenhouse gas emisions and emission reduction in an international context', *Institute for Environmental Studies*, Report R95/03. Vrije Universiteit Amsterdam, The Netherlands.

17 Mori, S (1995) 'A long-term evaluation of nuclear power technology by extended DICE + e model simulations – Multiregional approach for resources and industry allocation', *Progress in Nuclear Energy* 29: pp 135–142.

18 Manne, A S, Mendelsohn, R and Richels, R G (1993) 'MERGE: A model for evaluating regional and global effects of GHG reduction policies', *Energy Policy* 23: pp 17–34.

19 Cohan, D, Stafford, R K, Scheraga, J D and Herrod, S (1994) *The Global Climate Policy Evaluation Framework.* Proceedings of the 1994 A&WMA Global Climate Change Conference, Phoenix, 5–8 April, Air and Waste Management Association, Pittsburgh.

20 Nordhaus, W D and Yang, Z (1995) *RICE: A Regional Dynamic General Equilibrium Model of Optimal Climate Change Policy.* Yale University Press, New Haven, CT.

21 Kolstad, C D (1994) 'The timing of CO_2 control in the face of uncertainty and learning' in Van Ierland, E C (ed) *International Environmental Economics.* Elsevier, Amsterdam, pp 75–91

22 Weyant et al (1996) *Op cit*, Chapter 5, Note 40.

APPENDIX 2

ACCIS (1990) *Directory of United Nations Databases and Information Services* 4th Edition. United Nations, New York.

Beier, J (1991) *Global Change Data Sets: Excerpts from the Master Directory.* National Aeronautic and Space Administration (NASA), USA.

Burtis, M D (1992) *Catalog of Data Bases and Reports* ORNL/CDIAC-34/R4, Carbon Dioxide Information and Analysis Center, Oak Ridge National Laboratory, Oak Ridge, USA.

CIA (1972) *World Databank II.* Central Intelligence Agency, Washington, DC, USA.

ECMWF (1993) *The Description of the ECMWF/WCRP Level III-A Global Atmospheric Data Archive.* European Centre for Medium Range Weather Forecasting, Shinfield, Reading, UK.

FAO (1992) *AGROSTAT-PC, Computerized Information Series: User Manual, Population, Land Use, Production, Trade, Food Balance Sheets, Forest Products.* Food and Agriculture Organization, Rome.

FAO (1993) *Agriculture: Towards 2010* Report C93/24. Food and Agriculture Organization, Rome, 320 pp plus appendices.

Fritz, J (1990) *A Survey of Environmental Monitoring and Information Management Programmes of International Organizations.* United Nations Environment Programme, Nairobi.

Haxby, W F et al (1983) 'Digital images of combined oceanic and continental data sets and their use in tectonic studies', *EOS Transactions of the American Geophysical Union* 64: pp 995–1004.

Hicks, A J (1993) *Directory of Organizations and Institutes Active in Environmental Monitoring.* United Nations Environment Programme, Nairobi.

Hulme, M (1994) 'Validation of large-scale precipitation fields in general circulation models' in Desbois, M and Dèsalmand, F (eds) *Global Precipitation and Climate Change.* NATO ASW Proceedings. Springer-Verlag, Berlin, pp 387–405.

Hulme, M, Conway, D, Jones, P D, Jiang, T, Barrow, E M and Turney, C (1995a) 'Construction of a 1961–90 European climatology for climate change modelling and impact applications', *International Journal of Climatology* 15: pp 1333–1363.

Hulme, M, Jiang, T and Wigley, T M L (1995b) *SCENGEN, A Climate Change Scenario Generator: A User Manual*. Climatic Research Unit, University of East Anglia, Norwich, UK, 3 8pp.

Jones, P D (1994) 'Hemispheric surface air temperature variability – a reanalysis and an update to 1993', *Journal of Climate* 7: pp 1794–1802.

Leemans, R and Cramer, W P (1990) *The IIASA database for mean monthly values of temperature, precipitation, and cloudiness on a global terrestrial grid*. IIASA Research Report, RR-91-18. International Institute for Applied Systems Analysis, Laxenburg, Austria, 62 pp.

Leggett, J, Pepper, W J and Swart, R J (1992) 'Emissions scenarios for the IPCC: an update' in Houghton, J T, Callander, B A and Varney S K (eds) *Climate Change 1992: The Supplementary Report to the IPCC Scientific Assessment*. Cambridge University Press, pp 69–95.

Matthews, E (1983) 'Global vegetation and land use: new high-resolution data bases for climate studies', *Journal of Climate and Applied Meteorology*, 22: pp 473–487.

Matthews, E (1985) *Atlas of archived vegetation, land-use and seasonal albedo data sets*. NASA Technical Memorandum No 86199. National Aeronautics and Space Administration, New York, 23 pp.

Matthews, E and Fung, I (1987) 'Methane emission from natural wetlands' *Global Biogeochemical Cycles*, 1: pp 61–86.

Olson, J, Watts, J A and Allison, L J (1985) *Major World Ecosystem Complexes Ranked by Carbon in Live Vegetation: A Database*. Carbon Dioxide Information and Analysis Center, Oak Ridge, Tennessee, USA.

Tarpley, J D (1991) 'The NOAA global vegetation index project – a review', *Palaeoeography, Paleoclimatology and Paleoecology* 90: pp 189–194.

United Nations (1991) *World Population Prospects, 1990*. Population Studies No 120, United Nations, New York.

United Nations (1992) *World Population Prospects: The 1992 Revision*. United Nations, New York.

UNEP (1987) *World Directory of Environmental Expertise, INFOTERRA*. United Nations Environment Programme, Nairobi.

UNEP (1992) *World Atlas of Desertification*. Edward Arnold, London.

Viner, D and Hulme, M (1994) *The Climate Impacts LINK Project: Providing Climate Change Scenarios for Impacts Assessment in the UK*. Report for the UK Department of the Environment, Climatic Research Unit, Norwich, 24 pp.

Warrick, R A, Barrow, E M and Wigley, T M L (eds) (1993) *Climate and Sea Level Change: Observations, Projections and Implications*. Cambridge University Press, Cambridge.

WHO (1990) *Global estimates for health situation assessments and projections 1990*. WHO/HST/90.2. Division of Epidemiological Surveillance and Health Situation and Trend Assessment, World Health Organization, Geneva, 61 pp.

Wigley, T M L and Raper, S C B (1992) 'Implications for climate and sea level of revised IPCC emissions scenarios', *Nature* 357: pp 293–300.

World Bank (1991) *World Development Report, 1991*. Oxford University Press, New York.

WRI (1992) *World Resources, 1992–1993. A Guide to the Global Environment Toward Sustainable Development*. World Resources Institute, Oxford University Press, New York, 385 pp.

Zobler, L (1986) *A world soil file for global climate modelling* NASA Technical Memorandum 87802. National Aeronautics and Space Administration, New York.

INDEX

Printed and bound by CPI Group (UK) Ltd, Croydon, CR0 4YY

23/10/2024

01777678-0001